GCSE
Core Science

Complete Revision and Practice

Contents

How Science Works

Theories Come, Theories Go 1
Bias and How to Spot It 3
Science Has Limits 4

Section One — Nerves and Hormones

The Nervous System 6
Reflexes 7
The Eye 9
    Warm-Up and Exam Questions 11
Hormones 12
Hormonal and Nervous Responses 13
Puberty and the Menstrual Cycle 14
Controlling Fertility 16
    Warm-Up and Exam Questions 17
Homeostasis 18
Insulin 20
Insulin and Diabetes 21
    Warm-Up and Exam Questions 22
Revision Summary for Section One 23

Section Two — Diet and Health

Respiration 24
Respiration and Blood 25
    Warm-Up and Exam Questions 26
Digestion 27
    Warm-Up and Exam Questions 29
Diet and Exercise 30
Diet Problems 32
    Warm-Up and Exam Questions 34
Health Claims 35
Drugs 37
Medical Drugs 39
    Warm-Up and Exam Questions 41
Smoking and Alcohol 42
Solvents and Painkillers 44
    Warm-Up and Exam Questions 46
Causes of Disease 47
The Body's Defence Systems 48
Immunisation 49
Treating Disease — Past and Future 51
    Warm-Up and Exam Questions 53
    Exam Questions 54
Revision Summary for Section Two 55

Section Three — Genetics and Evolution

Variation in Plants and Animals 56
DNA and Genes 58
Asexual Reproduction 59
Sexual Reproduction and Variation 60
    Warm-Up and Exam Questions 62
    Exam Questions 63
Genetic Diagrams 64
Genetic Disorders 66
    Warm-Up and Exam Questions 68
    Exam Questions 69
Genetic Engineering 70
The Human Genome Project 72
Cloning 74
    Warm-Up and Exam Questions 76
    Exam Questions 77

Photosynthesis 78
    Warm-Up and Exam Questions 80
Ecosystems 81
Ecosystems and Species 82
Pyramids of Biomass and Food Chains 83
Competition and Populations 85
    Warm-Up and Exam Questions 87
    Exam Questions 88
Fossils 89
Evolution 91
Natural Selection 92
Adaptation 94
    Warm-Up and Exam Questions 96
Revision Summary for Section Three 97

Section Four — Atoms, Elements and Compounds

Atoms 98
Elements, Compounds and Mixtures 99
The Periodic Table 100
Balancing Equations 102
    Warm-Up and Exam Questions 103
    Exam Questions 104
Group 1 — The Alkali Metals 105
Group 7 — The Halogens 106
Group 0 — The Noble Gases 107
    Warm-Up and Exam Questions 108
Properties of Metals 109
Extraction of Metals 111
Extracting Pure Copper 113
Other Metals 114
    Warm-Up and Exam Questions 116
    Exam Questions 117
Revision Summary for Section Four 118

Section Five — The Earth

Limestone 119
Useful Products from Air and Salt 121
    Warm-Up and Exam Questions 123
Fractional Distillation of Crude Oil 124
Alkanes 125
Alkenes 126
Cracking Crude Oil 127
Burning Hydrocarbons 128
Using Crude Oil as a Fuel 130
    Warm-Up and Exam Questions 132
    Exam Questions 133
The Earth's Structure 134
Evidence for Plate Tectonics 135
Evolution of the Atmosphere 136
Changes in the Atmosphere 137
    Warm-Up and Exam Questions 139
    Exam Questions 140
Human Impact on the Environment 141
Global Warming 143
The Carbon Cycle 144
Air Pollution 145
Protecting the Atmosphere 147
    Warm-Up and Exam Questions 149
    Exam Questions 150
Biodiversity and Indicator Species 151
Recycling materials 152
    Warm-Up and Exam Questions 154
Revision Summary for Section Five 155

Contents

Section Six — Materials and Reactions

Polymers 156
Uses of Polymers 157
    Warm-Up and Exam Questions 159
Paints and Pigments 160
Perfumes 161
Fancy Materials 163
    Warm-Up and Exam Questions 165
Chemicals and Food 166
Food Additives 168
Plant Oils in Food 170
    Warm-Up and Exam Questions 172
Plant Oils as Fuel 173
Ethanol 174
    Warm-Up and Exam Questions 176
Hydration and Dehydration 177
Thermal Decomposition 178
Neutralisation and Oxidation 179
Chemical Tests 180
    Warm-Up and Exam Questions 182
    Exam Questions 183
Energy Transfer in Reactions 184
Kinetic Theory & Forces Between Particles 186
Chemical Reaction Rates 187
Collision Theory 189
    Warm-Up and Exam Questions 190
    Exam Questions 191
Revision Summary for Section Six 192

Section Seven — Heat and Energy

Moving and Storing Heat 193
Melting and Boiling 195
    Warm-Up and Exam Questions 197
Conduction 198
Convection 199
Heat Radiation 200
Saving Energy 202
    Warm-Up and Exam Questions 203
Energy 204
Efficiency 205
    Warm-Up and Exam Questions 207
Energy Sources 208
Nuclear Energy 210
Nuclear and Geothermal Energy 211
Wind and Solar Energy 212
Solar Energy 213
Biomass 214
Wave Energy 215
Hydroelectric Power 216
Pumped Storage and Power Stations 217
    Warm-Up and Exam Questions 218
    Exam Questions 219
Revision Summary for Section Seven 220

Section Eight — Electricity and Waves

Electric Current 221
Current, Voltage and Resistance 222
The Dynamo Effect 224
Power Stations & The National Grid 226
    Warm-Up and Exam Questions 228
    Exam Questions 229
Electrical Power 230
Electrical Safety Devices 232
    Warm-Up and Exam Questions 234
Waves — The Basics 235
Wave Behaviour 237
    Warm-Up and Exam Questions 239
Dangers of EM Radiation 240
    Warm-Up and Exam Questions 242
Refraction 243
Radio Waves 245
Microwaves and X-rays 246
More Uses of Waves 247
    Warm-Up and Exam Questions 249
    Exam Questions 250
Digital and Analogue Signals 251
Seismic Waves 253
    Warm-Up and Exam Questions 255
Revision Summary for Section Eight 256

Section Nine — Radioactivity and Space

Radioactivity 257
Three Kinds of Radioactivity 258
Half-Life 260
    Warm-Up and Exam Questions 261
Dangers from Nuclear Radiation 262
Uses of Nuclear Radiation 264
    Warm-Up and Exam Questions 266
The Solar System 267
Magnetic Fields and Solar Flares 269
Beyond the Solar System 271
    Warm-Up and Exam Questions 272
The Life Cycle of Stars 273
The Origins of the Universe 274
    Warm-Up and Exam Questions 276
Gravity, Mass and Weight 277
Exploring the Solar System 279
    Warm-Up and Exam Questions 281
Revision Summary for Section 9 282

Exam Skills

Thinking in Exams 283
Answering Experiment Questions 284

Answers 286

Index 302

The Periodic Table

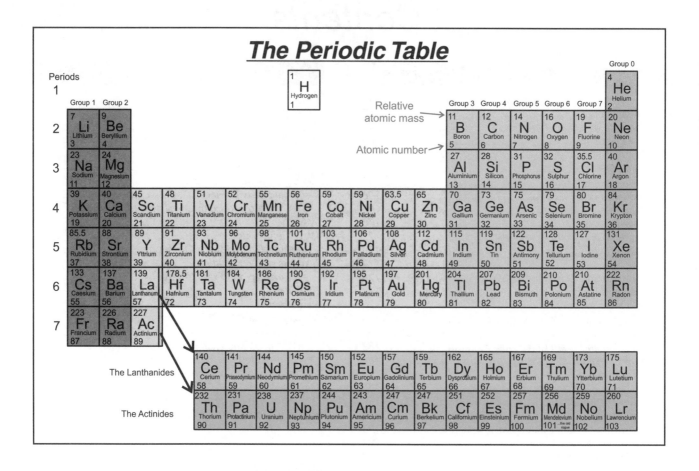

Published by Coordination Group Publications Ltd

Editors:
Amy Boutal, Ellen Bowness, Tom Cain, Katherine Craig, Gemma Hallam, Sarah Hilton,
Sharon Keeley, Andy Park, Rose Parkin, Kate Redmond, Alan Rix, Ami Snelling,
Laurence Stamford, Claire Thompson, Julie Wakeling, Sarah Williams.

Contributors:
Mike Bossart, Mark A. Edwards, Max Fishel, Sandy Gardner, Derek Harvey, Barbara Mascetti,
John Myers, Richard Parsons, Philip Rushworth, Adrian Schmit, Moira Steven,
Mike Thompson, Jim Wilson.

*With thanks to Science Photo Library for permission to reproduce the photographs
used on pages 202 and 270.*

With thanks to Jeremy Cooper, Ian Francis and Sue Hocking for the proofreading.

With thanks to Katie Steele for the copyright research.

ISBN: 978 1 84146 637 8

Groovy website: www.cgpbooks.co.uk

Printed by Elanders Hindson Ltd, Newcastle upon Tyne.
Jolly bits of clipart from CorelDRAW®

Theories Come, Theories Go

SCIENTISTS ARE ALWAYS RIGHT — OR ARE THEY?

Well it'd be nice if that were so, but it just ain't — never has been and never will be.
Increasing scientific knowledge involves making mistakes along the way. Let me explain...

Scientists come up with **hypotheses** — then **test** them

1) Scientists try and <u>explain</u> things. Everything.

2) They start by <u>observing</u> or <u>thinking about</u> something they don't understand — it could be anything,
   e.g. planets in the sky, a person suffering from an illness, what matter is made of... anything.

3) Then, using what they already know (plus a bit of insight),
   they come up with a <u>hypothesis</u> (a <u>theory</u>) that could <u>explain</u>
   what they've observed.

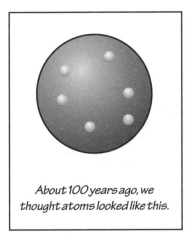

About 100 years ago, we thought atoms looked like this.

> Remember, a hypothesis is just a <u>theory</u>, a
> <u>belief</u>. And <u>believing</u> something is true doesn't
> <u>make</u> it true — not even if you're a scientist.

4) So the next step is to try and convince other scientists that
   the hypothesis is right — which involves using <u>evidence</u>.
   First, the hypothesis has to fit the <u>evidence</u> already available
   — if it doesn't, it'll convince <u>no one</u>.

5) Next, the scientist might use the hypothesis to make a <u>prediction</u> — a crucial step. If the hypothesis
   predicts something, and then <u>evidence</u> from <u>experiments</u> backs that up, that's pretty convincing.

> This <u>doesn't</u> mean the hypothesis is <u>true</u> (the 2nd prediction, or the
> 3rd, 4th or 25th one might turn out to be <u>wrong</u>) — but a hypothesis
> that correctly predicts something in the <u>future</u> deserves respect.

A hypothesis is a good place to start

You might have thought that science was all about facts... well, it's not as cut and dried as
that — you also need to know about the process that theories go through to become accepted,
and how those theories change over time. Remember, nothing is set in stone...

Theories Come, Theories Go

*Other scientists will **test** the hypotheses too*

1) Now then... <u>other</u> scientists will want to use the hypothesis to make their <u>own predictions</u>, and they'll carry out their <u>own experiments</u>. (They'll also try to <u>reproduce</u> earlier results.) And if all the experiments in all the world back up the hypothesis, then scientists start to have a lot of <u>faith</u> in it.

2) However, if a scientist somewhere in the world does an experiment that <u>doesn't</u> fit with the hypothesis (and other scientists can <u>reproduce</u> these results), then the hypothesis is in trouble. When this happens, scientists have to come up with a new hypothesis (maybe a <u>modification</u> of the old theory, or maybe a completely <u>new</u> one).

3) This process of testing a hypothesis to destruction is a vital part of the scientific process. Without the '<u>healthy scepticism</u>' of scientists everywhere, we'd still believe the first theories that people came up with — like thunder being the belchings of an angered god (or whatever).

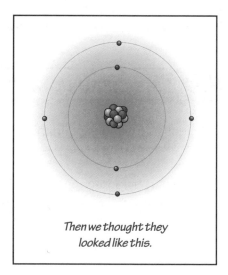

Then we thought they looked like this.

*If **evidence** supports a hypothesis, it's **accepted** — **for now***

1) If pretty much every scientist in the world believes a hypothesis to be true because experiments back it up, then it usually goes in the <u>textbooks</u> for students to learn.

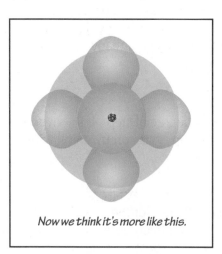

Now we think it's more like this.

2) Our <u>currently accepted</u> theories are the ones that have survived this 'trial by evidence' — they've been tested many, many times over the years and survived (while the less good ones have been ditched).

3) However... they never, <u>never</u> become hard and fast, totally indisputable <u>fact</u>.

> You can never know... it'd only take <u>one</u> odd, totally inexplicable result, and the hypothesising and testing would start all over again.

You expect me to believe that — then show me the evidence...

If scientists think something is true, they need to produce evidence to convince others — it's all part of <u>testing a hypothesis</u>. One hypothesis might survive these tests, while others won't — it's how things progress. And along the way some hypotheses will be disproved — i.e. shown not to be true. So, you see... not everything scientists say is true. <u>It's how science works</u>.

Bias and How to Spot it

Scientific results are often used to make a point, but results are sometimes presented in a biased way.

You don't need to *lie* to make things **biased**

1) For something to be misleading, it doesn't have to be untrue. We tend to read scientific facts and assume that they're the 'truth', but there are many different sides to the truth. Look at this headline...

> **1 in 2 people are of above average weight** *Sounds like we're a nation of fatties.*

2) But an average is a kind of 'middle value' of all your data. Some readings are higher than average (about half of them, usually). Others will be lower than average (the other half).

So the above headline could just as accurately say: **1 in 2 people are of below average weight**

3) The point is... both headlines sound quite worrying, even though they're not. That's the thing... you can easily make something sound really good or really bad — even if it isn't. You can...

---

① ...use only some of the data, rather than all of it:

"Many people lost weight using the new SlimAway diet. Buy it now!!"

"Many" could mean anything — e.g. 50 out of 5000 (i.e. 1%). But that could be ignoring most of the data.

② ...phrase things in a 'leading' way:

90% fat free!

Would you buy it if it were "90% cyanide free"? That 10% is the important bit, probably.

③ ...use a statistic that supports your point of view:

The amount of energy wasted is increasing. | Energy wasted per person is decreasing. | The rate at which energy waste is increasing is slowing down.

These describe the same data. But two sound positive and one negative.

Think about **why** things **might** be **biased**

1) People who want to make a point can sometimes present data in a biased way to suit their own purposes (sometimes without knowing they're doing it).

2) And there are all sorts of reasons why people might want to do this — for example...

> • Governments might want to persuade voters, other governments, journalists, etc. Evidence might be ignored if it could create political problems, or emphasised if it helps their cause.
> • Companies might want to 'big up' their products. Or make impressive safety claims, maybe.
> • Environmental campaigners might want to persuade people to behave differently.

3) People do it all the time. This is why any scientific evidence has to be looked at carefully. Are there any reasons for thinking the evidence is biased in some way?

> • Does the experimenter (or the person writing about it) stand to gain (or lose) anything?
> • Might someone have ignored some of the data for political or commercial reasons?
> • Is someone using their reputation rather than evidence to help make their case?

Scientific data's not always misleading, you just need to be careful. The most credible argument will be the one that describes all the data that was found, and gives the most balanced view of it.

Science Has Limits

Science can give us amazing things — cures for diseases, space travel, heated toilet seats...
But science has its limitations — there are questions that it just can't answer.

Some questions are **unanswered** by science — so far

1) We don't understand everything. And we never will. We'll find out more, for sure
— as more hypotheses are suggested, and more experiments are done.
But there'll always be stuff we don't know.

> For example, today we don't know as much as we'd like about
> climate change (global warming). Is climate change definitely
> happening? And to what extent is it caused by humans?

2) These are complicated questions, and at the moment scientists don't all agree on the answers.
But eventually, we probably will be able to answer these questions once and for all.

3) But by then there'll be loads of new questions to answer.

Other questions are **unanswerable** by science

1) Then there's the other type... questions that all the experiments in the world won't
help us answer — the "Should we be doing this at all?" type questions.
There are always two sides...

> The question of whether something is morally or ethically right
> or wrong can't be answered by more experiments — there is
> no "right" or "wrong" answer.

2) The best we can do is get a consensus from society — a judgement that most people are
more or less happy to live by. Science can provide more information to help people
make this judgement, and the judgement might change over time. But in the end it's up
to people and their conscience.

To answer or not to answer, that is the question...

It's official — no one knows everything. Your teacher/mum/annoying older sister (delete as applicable)
might think and act as if they know it all, but sadly they don't. So in reality you know one thing they
don't — which clearly makes you more intelligent and generally far superior in every way. Possibly.

Science Has Limits

*People have **different opinions** about **ethical questions***

1) Take <u>embryo screening</u> (which allows you to choose an embryo with particular characteristics). It's <u>possible</u> to do it — but does that mean we <u>should</u>?

2) Different people have <u>different opinions</u>. For example...

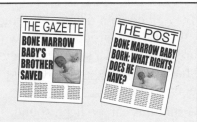

- Some people say it's <u>good</u>... couples whose <u>existing</u> child needs a <u>bone marrow transplant</u>, but who can't find a donor, will be able to have <u>another</u> child selected for its <u>matching</u> bone marrow. This would <u>save</u> the life of their first child — and if they <u>want</u> another child anyway... where's the harm?

- Other people say it's <u>bad</u>... they say it could have serious effects on the <u>child</u>. In the above example the new child might feel <u>unwanted</u> — thinking they were only brought into the world to help <u>someone else</u>. And would they have the right to <u>refuse</u> to donate their bone marrow (as anyone else would)?

*Loads of other **factors** can **influence decisions** too*

Here are some other factors that can influence decisions about science, and the way science is used:

Economic factors

- <u>Companies</u> very often won't pay for research unless there's likely to be a <u>profit</u> in it.

- Society can't always <u>afford</u> to do things scientists recommend without <u>cutting back elsewhere</u> (e.g. investing heavily in alternative energy sources).

Social factors

- Decisions based on scientific evidence affect <u>people</u> — e.g. should fossil fuels be taxed more highly (to invest in alternative energy)? Should alcohol be banned (to prevent health problems)? <u>Would the effect on people's lifestyles be acceptable...</u>

Environmental factors

- Genetically modified crops may help us produce more food — but some people say they could cause <u>environmental problems</u> (see page 70).

Science is a "real-world" subject...

Science isn't just done by people in white coats in labs who have no effect on the outside world. Science has a massive effect on the real world every day, and so real-life things like <u>money</u>, <u>morals</u> and <u>how people might react</u> need to be considered. It's why a lot of issues are so difficult to solve.

The Nervous System

The nervous system is what lets you react to what goes on around you, so you'd find life tough without it.

Sense organs detect stimuli

A stimulus is a change in your environment which you may need to react to (e.g. a recently pounced tiger). You need to be constantly monitoring what's going on so you can respond if you need to.

1) You have five different sense organs — eyes, ears, nose, tongue and skin.

2) They all contain different receptors. Receptors are groups of cells which are sensitive to a stimulus. They change stimulus energy (e.g. light energy) into electrical impulses.

3) A stimulus can be light, sound, touch, pressure, pain, chemical, or a change in position or temperature.

Sense organs and receptors
Don't get them mixed up:

The eye is a sense organ — it contains light receptors.

The ear is a sense organ — it contains sound receptors.

The five sense organs and the receptors that each contains:

1) Eyes   Light receptors.

2) Ears   Sound and "balance" receptors.

3) Nose   Smell receptors — sensitive to chemical stimuli.

4) Tongue   Taste receptors: — sensitive to bitter, salt, sweet and sour, plus the taste of savoury things like monosodium glutamate (MSG) — chemical stimuli.

5) Skin

Sensitive to touch, pressure and temperature change.

Sensory neurones

The nerve cells that carry signals as electrical impulses from the receptors in the sense organs to the central nervous system.

The central nervous system coordinates a response

1) The central nervous system (CNS) is where all the information from the sense organs is sent, and where reflexes and actions are coordinated.

The central nervous system consists of the brain and spinal cord only.

2) Neurones (nerve cells) transmit the information (as electrical impulses) very quickly to and from the CNS.

3) 'Instructions' from the CNS are sent to the effectors (muscles and glands), which respond accordingly.

Motor neurones

The nerve cells that carry signals to the effector muscles or glands.

Effectors

Muscles and glands are known as effectors. They respond in different ways — muscles contract in response to a nervous impulse, whereas glands secrete hormones.

Nervous System = 5 sense organs + neurones + brain + spinal cord

In the exam, you might have to take what you know about a human and apply it to a horse (easy... sound receptors in its ears, light receptors in its eyes, etc.), or to a snake (so if you're told that certain types of snakes have heat receptors in nostril-like pits on their head, you should know what type of stimulus those pits are sensitive to).

Reflexes

Your brain can <u>decide</u> how to respond to a stimulus <u>pretty quickly</u>.
But sometimes waiting for your brain to make a decision is just <u>too slow</u>. That's why you have <u>reflexes</u>.

Reflexes help prevent *injury*

1) <u>Reflexes</u> are <u>automatic</u> responses to certain stimuli — they can reduce the chances of being injured.

2) For example, if someone shines a <u>bright light</u> in your eyes, your <u>pupils</u> automatically get smaller so that less light gets into the eye — this stops it getting <u>damaged</u>.

3) Or if you get a shock, your body releases the <u>hormone</u> adrenaline automatically — it doesn't wait for you to <u>decide</u> that you're shocked.

4) The route taken by the information in a reflex (from receptor to effector) is called a <u>reflex arc</u>.

The **reflex arc** goes through the **central nervous system**

1) The neurones in many reflex arcs only go through the <u>spinal cord</u> — they <u>don't go to the brain</u>.

2) When a <u>stimulus</u> (e.g. a painful bee sting) is detected by receptors, an impulse is sent along a <u>sensory neurone</u> to the spinal cord.

3) In the spinal cord the sensory neurone passes on the message to another type of neurone — a <u>relay neurone</u>.

4) The relay neurone <u>relays</u> the impulse to a <u>motor neurone</u>.

5) The impulse then travels along the motor neurone to the <u>effector</u> (in the example below it's a muscle).

6) The muscle responds by <u>contracting</u>. A gland responds by <u>secreting</u>.

7) Because you don't have to think about the response (which takes time) it's <u>quicker</u> than normal responses.

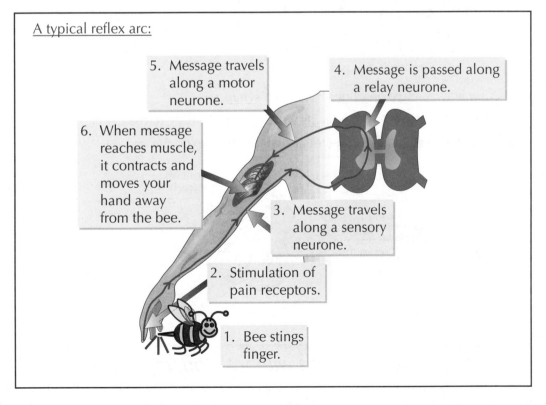

A typical reflex arc:

5. Message travels along a motor neurone.

4. Message is passed along a relay neurone.

6. When message reaches muscle, it contracts and moves your hand away from the bee.

3. Message travels along a sensory neurone.

2. Stimulation of pain receptors.

1. Bee stings finger.

Reflexes

Make sure you've learnt the order of a reflex arc.

Stimulus, receptor, neurones, effector, response

Here's a block diagram of a <u>reflex arc</u> — it shows what happens, from stimulus to response.

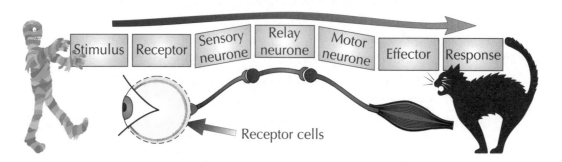

Stimulus | Receptor | Sensory neurone | Relay neurone | Motor neurone | Effector | Response

Receptor cells

Synapses connect neurones

1) <u>Neurones</u> (nerve cells) transmit information as <u>electrical impulses</u> around the body.

2) Neurones have <u>branched endings</u> so they can <u>connect</u> with lots of other neurones. And they're <u>long</u>, which also <u>speeds up</u> the impulse (fewer connections means a quicker signal).

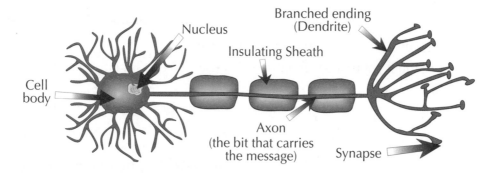

Nucleus

Branched ending (Dendrite)

Insulating Sheath

Cell body

Axon (the bit that carries the message)

Synapse

3) The connection between <u>two neurones</u> is called a <u>synapse</u>.

4) The nerve signal is transferred by <u>chemicals</u> which <u>diffuse</u> (move) across the gap.

5) These chemicals then set off a <u>new electrical signal</u> in the <u>next</u> neurone.

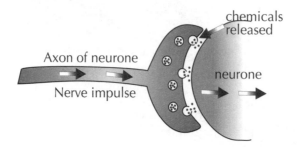

chemicals released

Axon of neurone

Nerve impulse

neurone

A reflex action lets you respond without stopping and thinking

The difference between a reflex and a "considered response" is the involvement of the conscious part of your brain. Reflexes may bypass your brain completely — your body just gets on with things.

The Eye

The eye's a good example of a sense organ...

Learn *the eye* with all its labels:

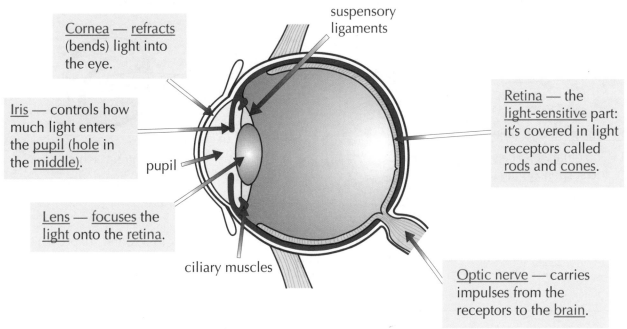

Cornea — refracts (bends) light into the eye.

Iris — controls how much light enters the pupil (hole in the middle).

Lens — focuses the light onto the retina.

ciliary muscles

suspensory ligaments

pupil

Retina — the light-sensitive part: it's covered in light receptors called rods and cones.

Optic nerve — carries impulses from the receptors to the brain.

Rods are more sensitive in dim light but can't sense colour. Cones are sensitive to colours but are not so good in dim light (red-green colour blindness is due to a lack of certain cone cells).

The *iris reflex* — adjusting for *bright light*

Very bright light can damage the retina — so you have a reflex to protect it.

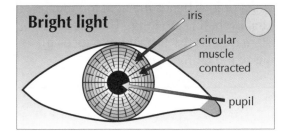

Bright light

iris

circular muscle contracted

pupil

1) Very bright light triggers a reflex that makes the pupil smaller, allowing less light in. (See p7 for more about reflexes... but basically, light receptors detect the bright light, send a message to an unconscious part of the brain along a sensory neurone, and then the brain sends a message straight back along a motor neurone telling the circular muscles in the iris to contract, which makes the pupil smaller.)

2) The opposite process happens in dim light. This time, the brain tells the radial muscles to contract, which makes the pupil bigger.

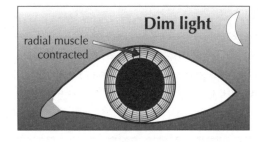

Dim light

radial muscle contracted

The Eye

This page is all about the ciliary muscles and their role in accommodation.
Not to be confused with the circular muscles and their role in the iris reflex...

Focusing on near and distant objects — another **reflex**

The eye focuses light by changing the <u>shape</u> of the <u>lens</u> — this is known as <u>accommodation</u>.

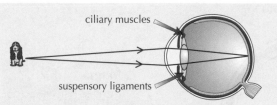

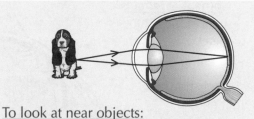

To look at distant objects:

1) The <u>ciliary muscles relax</u>, which allows the <u>suspensory ligaments</u> to <u>pull tight</u>.

2) This makes the lens go <u>thin</u>.

To look at near objects:

1) The <u>ciliary muscles contract</u>, which <u>slackens</u> the <u>suspensory ligaments</u>.

2) The lens becomes <u>fat</u>.

As you get older, your eye's lens loses <u>flexibility</u>, so it can't easily spring back to a round shape. This means light <u>can't</u> be focused well for near viewing, so older people often have to use <u>reading glasses</u>.

1) <u>Long-sighted people</u> are <u>unable to focus</u> on <u>near</u> objects. This occurs when the <u>cornea</u> or <u>lens</u> doesn't <u>bend</u> the light enough or the <u>eyeball</u> is too <u>short</u>. The images of near objects are brought into focus <u>behind</u> the <u>retina</u>.

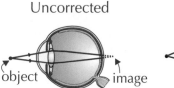

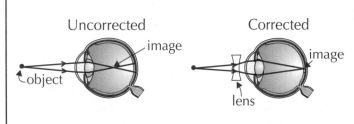

2) <u>Short-sighted people</u> are <u>unable to focus</u> on <u>distant</u> objects. This occurs when the <u>cornea</u> or <u>lens</u> bends the light <u>too much</u> or the <u>eyeball</u> is too <u>long</u>. The images of distant objects are brought into focus <u>in front</u> of the <u>retina</u>.

Binocular vision lets you judge **depth**

Some animals, including humans, have two eyes which <u>work together</u> — this is <u>binocular vision</u>. The brain uses small differences between what each eye sees to <u>judge distances</u> and <u>how fast</u> things are moving. It's handy for <u>catching prey</u> and deciding if it's safe to cross a road.

I think I'm a little long-sighted...

To see how important binocular vision is, cover one eye and try pouring water into a glass at arm's length. That's why you never see turkeys or horses pouring themselves a drink — no binocular vision.

Warm-Up and Exam Questions

Warm-up Questions

1) What are the five sense organs in the human body?
2) What is the role of the central nervous system (CNS)?
3) In what form is information transmitted along nerve cells?
4) The eye can focus on near objects and on distant objects by changing the shape of the lens. What is this known as?
5) Why is it an advantage to have binocular vision?
6) What name is given to the gap between two nerve cells?

Exam Questions

1   Gordon accidentally touches a hot object, causing his hand to immediately move away from it. The diagram below shows some of the parts involved in this response.

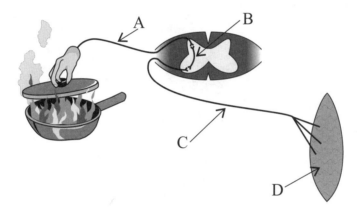

(a)   What is the name for this type of automatic response?

(1 mark)

(b)   On the diagram:

(i)   Which letter points to a relay neurone?

(1 mark)

(ii)   Which letter points to an effector?

(1 mark)

(c)   Explain how an electrical impulse in one neurone is able to pass to the next neurone.

(2 marks)

(d)   Give one physiological advantage, to the body, of these automatic responses.

(1 mark)

2   (a)   Explain how the iris reflex can reduce the amount of light entering the eye.

(4 marks)

(b)   What is the advantage of this reflex?

(1 mark)

Hormones

The other way to send information around the body (apart from along nerves) is by using hormones.

Hormones are chemical messengers sent in the blood

1) <u>Hormones</u> are <u>chemicals</u> released directly into the <u>blood</u>. They're carried in the <u>blood plasma</u> to other parts of the body, but only affect particular cells (called <u>target cells</u>) in particular places. Hormones control things in organs and cells that need <u>constant adjustment</u>.

2) Hormones are produced in various <u>glands</u>, as shown on the diagram. They travel through your body at "<u>the speed of blood</u>".

3) Hormones tend to have relatively <u>long-lasting</u> effects.

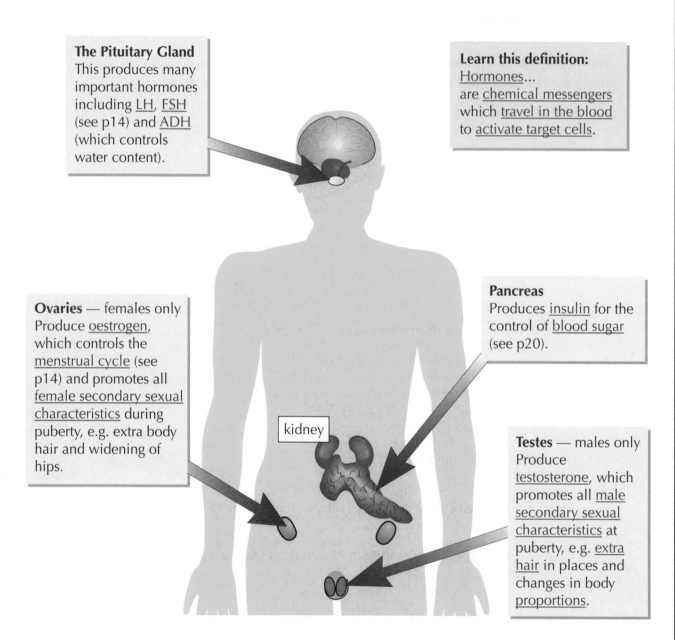

The Pituitary Gland
This produces many important hormones including <u>LH</u>, <u>FSH</u> (see p14) and <u>ADH</u> (which controls water content).

Learn this definition:
<u>Hormones</u>...
are <u>chemical messengers</u> which <u>travel in the blood</u> to <u>activate target cells</u>.

Ovaries — females only
Produce <u>oestrogen</u>, which controls the <u>menstrual cycle</u> (see p14) and promotes all <u>female secondary sexual characteristics</u> during puberty, e.g. extra body hair and widening of hips.

kidney

Pancreas
Produces <u>insulin</u> for the control of <u>blood sugar</u> (see p20).

Testes — males only
Produce <u>testosterone</u>, which promotes all <u>male secondary sexual characteristics</u> at puberty, e.g. <u>extra hair</u> in places and changes in body <u>proportions</u>.

These are just examples — there are loads more, each doing its own thing.

Hormonal and Nervous Responses

You'll be expected to be able to compare nervous responses with hormonal ones, so have a look at this...

Hormones and nerves do similar jobs, but there are differences

Nerves

1) Very <u>fast</u> message.

2) Act for a very <u>short time</u>.

3) Act on a very <u>precise</u> area.

4) <u>Electrical</u> message.

Hormones

1) <u>Slower</u> message.

2) Act for a <u>long time</u>.

3) Act in a more <u>general</u> way.

4) <u>Chemical</u> message.

If you're not sure whether a response is nervous or hormonal, have a think about the <u>three differences</u> — speed, longevity and affected area.

If the response is really quick, it's probably nervous

Some information needs to be passed to effectors <u>really quickly</u> (e.g. <u>pain</u> signals, or information from your <u>eyes</u> telling you about the <u>lion</u> heading your way), so it's no good using hormones to carry the message — they're <u>too slow</u>.

But if a response lasts for a long time, it's probably hormonal

For example, when you get a <u>shock</u>, a hormone called <u>adrenaline</u> is released into the bloodstream (causing the fight-or-flight response, where your body is hyped up ready for action). You can tell it's a hormonal response (even though it kicks in pretty quickly) because you feel a bit <u>wobbly</u> for a while <u>afterwards</u>.

Learn the differences between nervous and hormonal responses

Hormones control various organs and cells in the body, though they tend to control things that aren't immediately life-threatening. For example, they take care of most things to do with sexual development, pregnancy, birth, breast-feeding, blood sugar levels, water content... and so on.

Puberty and the Menstrual Cycle

Hormones control almost everything to do with <u>sex</u> and <u>reproduction</u>.

Hormones promote *sexual characteristics* at *puberty*

At puberty your body starts releasing <u>sex hormones</u> — <u>testosterone</u> in men and <u>oestrogen</u> in women. These trigger off the <u>secondary sexual characteristics</u>:

In men

1) <u>Extra hair</u> on face and body.
2) <u>Muscles develop</u>.
3) <u>Penis and testicles</u> enlarge.
4) <u>Sperm</u> production.
5) <u>Deepening</u> of <u>voice</u>.

In women

1) <u>Extra hair</u> on underarms and pubic area.
2) <u>Hips widen</u>.
3) Development of <u>breasts</u>.
4) <u>Egg release</u> and <u>periods start</u>.

The *menstrual cycle* has *four stages*

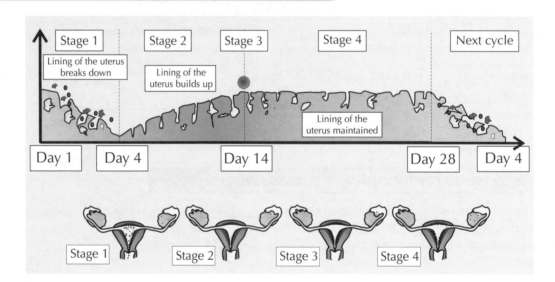

Stage 1

<u>Day 1 is when the bleeding starts</u>.
The uterus lining breaks down for about four days.

Stage 2

<u>The lining of the uterus builds up again</u>, from day 4 to day 14, into a thick spongy layer full of blood vessels, ready to receive a fertilised egg.

Stage 3

<u>An egg is developed and then released</u> from the ovary at day 14.

Stage 4

<u>The wall is then maintained</u> for about 14 days, until day 28. If no fertilised egg has landed on the uterus wall by day 28, the spongy lining starts to break down and the whole cycle starts again.

Puberty and the Menstrual Cycle

There are only a few hormones you need to know about in this section, and apart from insulin, they're all on this page...

The **menstrual cycle's** controlled by **four hormones**

1. FSH (follicle-stimulating hormone)

1) Causes an <u>egg to develop</u> in one of the ovaries.

2) Stimulates the <u>ovaries</u> to produce <u>oestrogen</u>.

Produced in the <u>pituitary gland</u>.

2. LH (luteinising hormone)

Stimulates the <u>release of an egg</u> at day 14.

Produced in the <u>pituitary gland</u>.

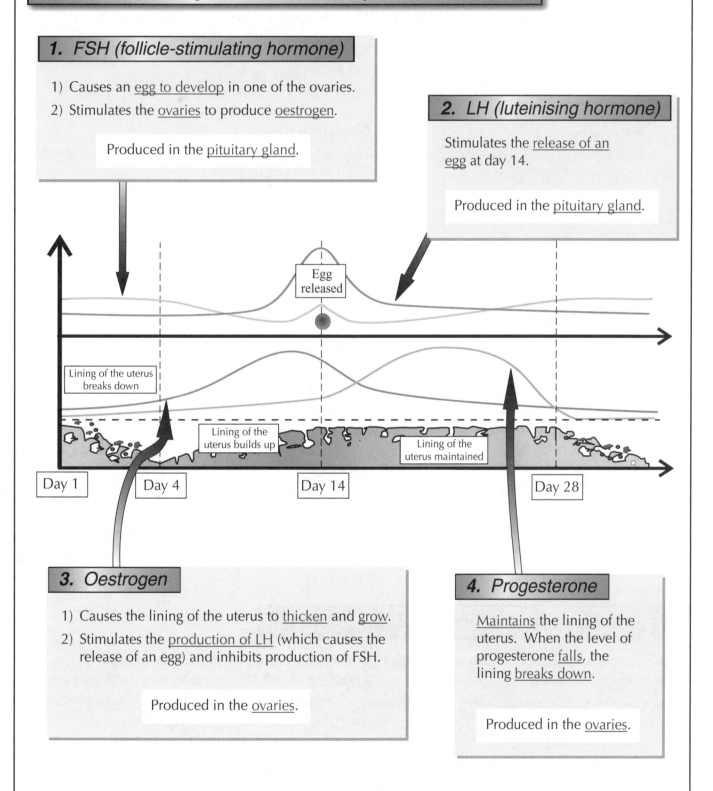

Egg released

Lining of the uterus breaks down

Lining of the uterus builds up

Lining of the uterus maintained

Day 1   Day 4   Day 14   Day 28

3. Oestrogen

1) Causes the lining of the uterus to <u>thicken</u> and <u>grow</u>.

2) Stimulates the <u>production of LH</u> (which causes the release of an egg) and inhibits production of FSH.

Produced in the <u>ovaries</u>.

4. Progesterone

<u>Maintains</u> the lining of the uterus. When the level of progesterone <u>falls</u>, the lining <u>breaks down</u>.

Produced in the <u>ovaries</u>.

OK, I admit it — this is quite hard to get your head around

In the exam, you might have to explain what hormone causes what in the menstrual cycle. It's tough, because they're all interlinked... but if you know your stuff, it should be quite straight forward.

Controlling Fertility

The hormones <u>FSH</u>, <u>oestrogen</u> and <u>LH</u> can be used to change artificially <u>how fertile</u> a woman is.

Hormones can be used to reduce fertility...

1) The hormone <u>oestrogen</u> can be used to <u>prevent</u> the <u>release</u> of an <u>egg</u> — so oestrogen can be used as a method of <u>contraception</u>. <u>The pill</u> is an oral contraceptive that contains oestrogen.

2) This may seem kind of strange (since naturally oestrogen helps stimulate the release of eggs). But if oestrogen is taken <u>every day</u> to keep the level of it <u>permanently high</u>, it <u>inhibits</u> the production of <u>FSH</u>, and <u>after a while</u> egg development and production <u>stop</u> and stay stopped.

Advantages

1) The pill's <u>over 99% effective</u> at preventing pregnancy.

2) It <u>reduces</u> the <u>risk</u> of getting some types of <u>cancer</u>.

Disadvantages

1) It <u>isn't 100% effective</u> — there's still a very slight chance of getting pregnant.

2) It can cause <u>side effects</u> like headaches, nausea, irregular menstrual bleeding, and fluid retention.

3) It <u>doesn't protect</u> against <u>sexually transmitted infections</u> (STIs).

...or increase it

1) Some women have levels of <u>FSH</u> that are <u>too low</u> to cause their <u>eggs to mature</u>. This means that <u>no eggs are released</u> and the women <u>can't get pregnant</u>.

2) The hormone FSH can be taken by these women to stimulate <u>egg production</u> in their <u>ovaries</u>. (Well, in reality... <u>FSH</u> stimulates the <u>ovaries</u> to produce <u>oestrogen</u>, which stimulates the pituitary gland to produce LH, which stimulates the <u>release of an egg</u>... but FSH has the desired effect anyway.)

Advantage

It helps a lot of women to <u>get pregnant</u> when previously they couldn't... pretty obvious.

Disadvantages

1) It <u>doesn't always work</u>.

2) <u>Too many eggs</u> could be stimulated, resulting in unexpected <u>multiple pregnancies</u> (twins, triplets etc.).

IVF can also help couples to have children

IVF ("in vitro fertilisation") involves collecting <u>eggs</u> from the woman's <u>ovaries</u> and <u>fertilising</u> them in a <u>lab</u> using the man's <u>sperm</u>. These are then grown into <u>embryos</u>, which are <u>transferred</u> to the woman's uterus.

1) <u>Hormones</u> are given <u>before</u> IVF to stimulate egg production (so <u>several</u> eggs can be collected).

2) <u>Oestrogen</u> and <u>progesterone</u> are often given to make <u>implantation</u> of the embryo into the uterus more likely to succeed.

But the use of hormones in IVF can cause problems for some women...

1) Some women experience <u>abdominal pain</u>, <u>vomiting</u> and <u>dehydration</u>.

2) There have been some reports of an <u>increased risk of cancer</u> due to the hormonal treatment (though others have reported <u>no such risk</u> — the position isn't really clear at the moment).

Warm-Up and Exam Questions

You could skim through this page in a few minutes, but there's no point unless you check over any bits you don't know and make sure you understand everything. It's not quick but it's the only way.

Warm-up Questions

1) Define the term hormone.
2) Give three differences between nervous and hormonal responses.
3) Which organ in the human body produces insulin?
4) List four male secondary sexual characteristics.
5) Briefly describe the process of in vitro fertilisation.

Exam Questions

1   Which of the following statements about the oestrogen-containing contraceptive pill is **not** true?

   A   It does not protect against sexually transmitted diseases.

   B   It is over 99% effective at preventing pregnancy.

   C   It is free from side effects.

   D   It reduces the risk of developing some forms of cancer.

   (1 mark)

2   Which of the following is controlled by the hormone FSH?

   A   Development of secondary sexual characteristics.

   B   Production of oestrogen by the ovaries.

   C   Release of LH from the pituitary.

   D   Release of an egg from the ovary.

   (1 mark)

3   The menstrual cycle is controlled by several different hormones.

   (a)   What effect does oestrogen have on the uterus lining?

   (1 mark)

   (b)   What is the effect of progesterone on the uterus lining?

   (1 mark)

   (c)   On what day of the menstrual cycle is the egg released?

   (1 mark)

4   Give **two** potential side-effects associated with IVF.

   (2 marks)

Homeostasis

Homeostasis involves balancing body functions to maintain a "constant internal environment". Hormones are sometimes (but not always) involved.

Homeostasis is maintaining a constant internal environment

Conditions in your body need to be kept steady so that cells can function properly. This involves balancing inputs (stuff going into your body) with outputs (stuff leaving). For example...

1) Levels of underline{CO_2} — respiration in cells (p24) constantly produces CO_2, which you need to get rid of.

2) Levels of oxygen — you need to replace the oxygen that your cells use up in respiration.

3) Water content — you need to keep a balance between the water you gain (in drink and food, and from respiration) and the water you pee, sweat and breathe out.

4) Body temperature — you need to get rid of excess body heat when you're hot, but retain heat when the environment is cold.

Negative feedback helps keep all these things steady

Changes in the environment trigger a response that counteracts the changes — e.g. a rise in body temperature causes a response that lowers body temperature.

This means that the internal environment tends to stay around a norm, the level at which the cells work best.

This only works within certain limits — if the environment changes too much then it might not be possible to counteract it (see the bit about heatstroke at the bottom of page 19).

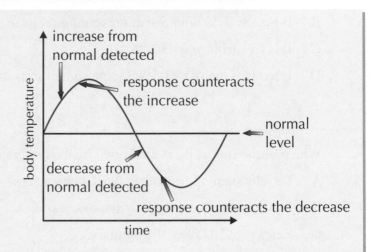

Water is lost from the body in various ways

Water is taken into the body as food and drink and is lost from the body in these ways:

1) through the skin as sweat...

2) via the lungs in breath...

3) via the kidneys as urine.

Some water is also lost in faeces.

The balance between sweat and urine can depend on what you're doing, or what the weather's like...

On a cold day, or when you're not exercising, you don't sweat much, so you'll produce more urine, which will be pale (since the waste carried in the urine is more diluted).

On a hot day, or when you're exercising, you sweat a lot, and so you will produce less urine, but this will be more concentrated (and hence a deeper colour). You will also lose more water through your breath when you exercise because you breathe faster.

Homeostasis

Body temperature is kept at about 37 °C

1) All <u>enzymes</u> work best at a certain temperature. The enzymes in the human body work best at about <u>37 °C</u> — and so this is the temperature your body tries to maintain.

2) A part of the <u>brain</u> acts as your own <u>personal thermostat</u>. It's sensitive to the blood temperature in the brain, and it <u>receives</u> messages from the skin that provide information about <u>skin temperature</u>.

3) To keep you at this temperature your body does these things:

When you're **too hot**:

1) <u>Hairs</u> lie flat.

2) <u>Lots of sweat</u> is produced — when it <u>evaporates</u> it <u>transfers heat</u> from you to the environment, cooling you down.

3) <u>Blood vessels</u> close to the surface of the skin <u>widen</u> (vasodilation). This allows more blood to flow near the surface, so it can radiate more heat into the surroundings.

4) You can <u>take off</u> layers of <u>clothing</u> to help you cool down.

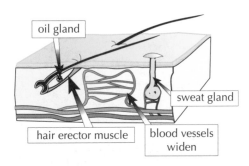

If you're exposed to <u>high temperatures</u> you can get <u>dehydrated</u> and you could get <u>heat stroke</u>. This can <u>kill</u> you (see below).

When you're **too cold**:

1) <u>Hairs</u> stand on end to trap an insulating layer of air which helps keep you warm.

2) <u>Very little sweat</u> is produced.

3) <u>Blood vessels</u> near the surface <u>constrict</u> (vasoconstriction) so that less heat can be transferred from the blood to the surroundings.

4) You <u>shiver</u>, and the movement generates heat in the muscles. <u>Exercise</u> does the same.

5) You can put on <u>more clothes</u>, to trap the heat in.

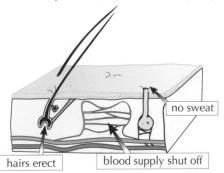

Your body temperature can drop to dangerous levels if you're exposed to <u>very low temperatures</u> for a long time — this is called <u>hypothermia</u>. If you don't get help quickly you can <u>die</u>.

Ion content is regulated by the kidneys

1) <u>Ions</u> (e.g. sodium, Na⁺) are taken into the body in <u>food</u>, then absorbed into the blood.

2) If the food contains <u>too much</u> of any kind of ion then the excess ions need to be <u>removed</u>. E.g. a salty meal will contain far too much Na⁺.

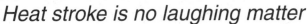
Kidneys

3) Some ions are lost in <u>sweat</u> (which tastes salty, you'll have noticed).

4) The kidneys will <u>remove the excess</u> from the blood — this is then got rid of in <u>urine</u>.

5) <u>Sports drinks</u> (which usually contain <u>electrolytes</u> and <u>carbohydrates</u>) can help your body keep things in order. The <u>electrolytes</u> (e.g. sodium) replace those lost in <u>sweat</u>, while the carbohydrates can give a bit of an energy boost. But claims about sports drinks need to be looked at carefully.

Heat stroke is no laughing matter

If you're in really high temperatures for a long time you can get heat stroke — sweating stops, since you get so dehydrated, and there's a big rise in your body temperature. If you don't cool down you can die.

Insulin

Insulin is a hormone which controls how much sugar there is in your blood.

Insulin controls blood sugar levels

1) Eating foods rich in <u>carbohydrates</u> puts a lot of <u>glucose</u> (a type of sugar) into the blood from the gut.

2) The normal metabolism of cells <u>removes</u> glucose from the blood.

3) Vigorous <u>exercise</u> removes much more glucose from the blood.

4) Obviously, to control the <u>level</u> of blood glucose there has to be a way to <u>add or remove</u> glucose from the blood. And this is it:

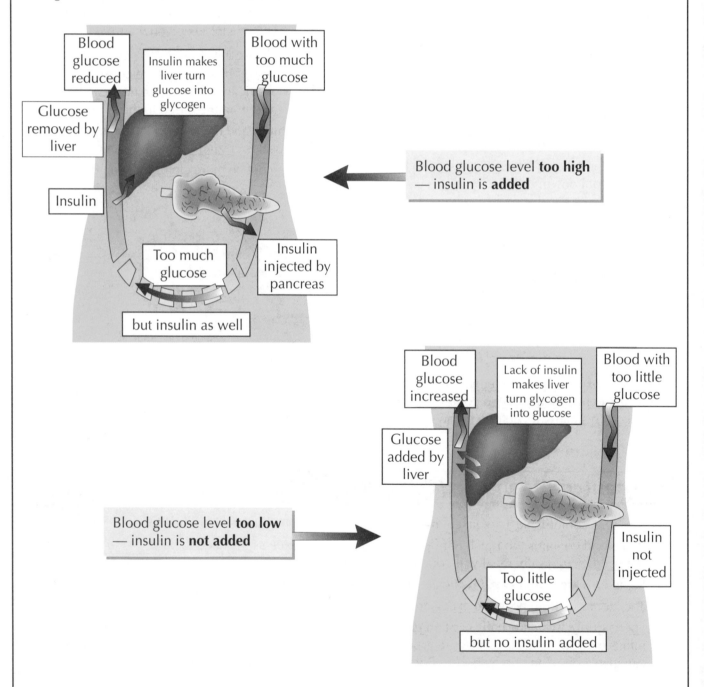

Remember, the liver stores energy as **glycogen**, not glucose

Glycogen's pretty similar to starch really — it's just a load of glucose molecules joined together. It's great for storage because it's too big to pass out of cells (unlike glucose, which would go everywhere...).

Insulin and Diabetes

People with diabetes have to make sure their blood sugar levels don't get too high.

Diabetes — the pancreas stops making enough insulin

1) Diabetes (type 1) is a disease in which the pancreas doesn't produce enough insulin. The result is that a person's blood sugar can rise to a level that can kill them.

2) The problem can be controlled in two ways:

1

Avoiding foods rich in carbohydrates (which turn to glucose when digested).

Remember, insulin reduces blood sugar levels.

2

1) Injecting insulin into the blood before meals (especially if the meal is high in carbohydrates).

2) Injecting insulin makes the liver remove the glucose from the blood as soon as it enters it from the gut, when the (carbohydrate-rich) food is being digested.

3) This stops the level of glucose in the blood from getting too high and is a very effective treatment.

Insulin is made using genetically engineered bacteria

1) Insulin used by diabetics used to come from cows and pigs. The cows and pigs were slaughtered and insulin was extracted from their pancreases and purified.

2) This can't produce enough insulin to keep up with the demand though. Also, the insulin isn't quite the same as the human form, which means the immune system often attacks it, reducing its efficiency.

3) So... bacteria are now genetically modified to include the human insulin gene (see page 70). They produce a pure form of human insulin which isn't rejected by the immune system — and they can produce as much as is needed.

It's important to learn how to control diabetes

Hormones control a lot of your body's functions. So if one of these functions isn't being done quite right, it might be possible to fix it by injecting suitable hormones — just like with diabetes.

Warm-Up and Exam Questions

Warm-up Questions

1) What is meant by the term homeostasis?
2) List six things that need to be kept constant within the body.
3) List four ways in which water is lost from the body.
4) Why is it important that human body temperature is kept at about 37 °C?
5) Give two advantages of using insulin from genetically engineered microorganisms to treat people with diabetes compared to using insulin obtained from pigs and cows?

Exam Questions

1   Describe the four ways in which the body responds to a drop in temperature.

(4 marks)

2   Which organ is responsible for regulating the ion content of the blood?

(1 mark)

3   The diagram below shows how blood sugar levels are regulated in humans.

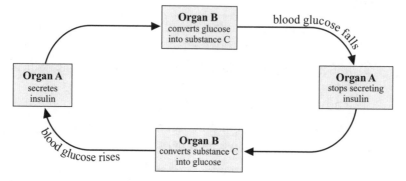

(a) What is the name for this type of control mechanism?

(1 mark)

(b) From the diagram, identify:

   (i)   organ A

(1 mark)

   (ii) organ B

(1 mark)

   (iii) substance C

(1 mark)

(c) (i)   Suggest a reason why blood glucose might rise.

(1 mark)

   (ii) What process constantly removes glucose from the blood?

(1 mark)

   (iii) What would cause blood glucose levels to fall rapidly?

(1 mark)

(d) Give two ways in which type 1 diabetes can be controlled.

(2 marks)

Revision Summary for Section One

Congratulations, you've made it to the end of the first section. I reckon that section wasn't too bad, there's some pretty interesting stuff there — nerves, hormones, homeostasis... what more could you want? Actually, I know what more you could want... some questions to make sure you know it all.

1) List the five sense organs and the receptors that each one contains.

2) Where would you find the following receptors in a dog:
   a) smell b) taste c) light d) pressure e) sound.

3) What is the purpose of a reflex action?

4) Describe the pathway of a reflex arc from stimulus to response.

5) Draw a diagram of a typical neurone, labelling all its parts.

6) What's a synapse? How are signals passed across a synapse?

7) Describe the iris reflex. Why is this important?

8) How does accommodation of the eye work? Is the lens fat or thin to look at distant objects?

9) Give two examples of hormones, saying where they're made and what they do.

10)*Here's a table of data about response times.
   a) Which response (A or B) is carried by nerves?
   b) Which is carried by hormones?

Response	Reaction time (s)	Response duration (s)
A	0.005	0.05
B	2	10

11) What secondary sexual characteristics does testosterone trigger in males?
    And oestrogen in females?

12) Draw a timeline of the 28-day menstrual cycle.
    Label the four stages of the cycle and label when the egg is released.

13) What roles do oestrogen, FSH and LH play in the menstrual cycle?

14) State two advantages and two disadvantages of using the contraceptive pill.

15) Which hormone is used to stimulate egg production in fertility treatment?

16) Describe some of the main issues in the IVF debate.

17) Explain how negative feedback helps to maintain a constant internal environment.

18) Describe how the amount and concentration of urine you produce varies depending on how much exercise you do and how hot it is.

19) Describe how body temperature is reduced when you're too hot. What happens if you're cold?

20) Briefly describe the relationship between food, exercise, blood glucose and insulin.

21) Define diabetes and explain two ways in which it can be controlled.

*Answers on page 286.

Respiration

Respiration does <u>NOT</u> mean breathing in and out.
<u>Respiration</u> actually goes on in <u>every cell</u> in your body.

Respiration *is the process of* releasing energy *from* glucose

1) Respiration is the process of <u>releasing energy</u> from <u>glucose</u>. This energy is used to do things like: build up <u>larger molecules</u> (e.g. proteins), contract <u>muscles</u>, maintain a steady <u>body temperature</u>...

2) There are <u>two types</u> of respiration, <u>aerobic</u> and <u>anaerobic</u>.

Aerobic respiration *needs plenty of* oxygen

1) <u>Aerobic respiration</u> happens when there's <u>plenty of oxygen</u> available. ("<u>Aerobic</u>" means "<u>with oxygen</u>".)

2) This is the most efficient way to release <u>energy</u> from <u>glucose</u>, and is the type of respiration that you use <u>most of the time</u>.

> **Glucose + Oxygen → Carbon dioxide + Water (+ ENERGY)**

3) You also need to know the <u>symbol equation</u>.

$$C_6H_{12}O_6 + 6O_2 \rightarrow 6CO_2 + 6H_2O \;\; (+ \text{ENERGY})$$

Anaerobic respiration *does* not *use oxygen at all*

1) When you do really <u>vigorous exercise</u>, your body can't supply enough <u>oxygen</u> to your muscles for aerobic respiration. Your muscles have to start <u>respiring anaerobically</u> as well.

2) "<u>Anaerobic</u>" just means "<u>without</u> oxygen". It's <u>NOT</u> the best way to convert glucose into energy because it releases much <u>less energy</u>, and <u>lactic acid</u> is also produced (which builds up in the muscles, making them <u>painful</u> and <u>fatigued</u>). But at least you can keep on using your muscles.

> **Glucose → Lactic Acid   (+ ENERGY)**

3) After resorting to anaerobic respiration, when you stop exercising you'll have an <u>oxygen debt</u>. You need <u>extra oxygen</u> to break down all the lactic acid that's built up in your muscles.

4) This means you have to keep <u>breathing hard</u> for a while <u>after you stop</u> exercising — to repay the debt. Your <u>heart rate</u> has to <u>stay high</u> too to carry the lactic acid to the liver to be broken down.

Unfit people have to resort to <u>anaerobic</u> respiration <u>quicker</u> than fit people do.

One big deep breath and learn it all

It's really not as complicated as it looks. Honest. And you need to remember those three equations. It's the equations that tell you what respiration really is — the release of energy from glucose. Just remember that fact when it comes to the exam and be sure not to say that respiration's breathing...

Respiration and Blood

So, every single cell in your body needs oxygen for aerobic respiration. That's where blood comes in...

The **oxygen** for respiration is carried by the **blood**

1) Blood is <u>pumped</u> around the body by the contractions of the <u>heart</u>.

2) The blood leaves the heart and flows through <u>arteries</u>. These split into thousands of tiny capillaries, which take blood to every cell in the body. The blood then flows back to the heart through veins.

Blood pressure (measured in millimetres of mercury, mmHg) is at its <u>highest</u> when the heart <u>contracts</u> — this is the <u>systolic pressure</u>. When the heart <u>relaxes</u>, the pressure is at its <u>lowest</u> — the <u>diastolic pressure</u>.

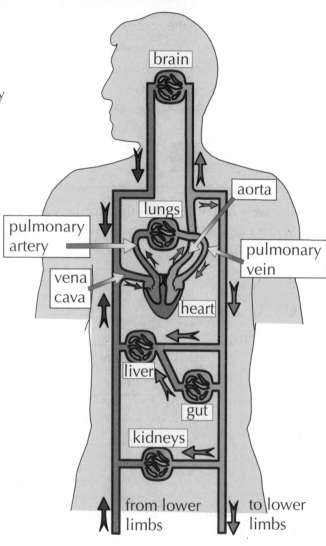

Red blood cells carry the oxygen

1) Oxygen is carried by red blood cells. They have a doughnut shape to give a <u>large surface area</u> for absorbing <u>oxygen</u>.

2) They contain a substance called <u>haemoglobin</u>. In the <u>lungs</u>, haemoglobin combines with <u>oxygen</u>. In body tissues the oxygen is released.

Red blood cells have a large surface area for absorbing oxygen

There's other stuff in your blood as well — e.g. white blood cells (which help fight disease — see p48), platelets (which help your blood clot), and plasma (the liquid that all the different bits and bobs float around in). But it's the red blood cells that carry the oxygen — remember that.

Warm-Up and Exam Questions

Take a deep breath and go through these warm-up questions one by one.
If you don't know the basic facts there's no way you'll cope with the exam questions.

Warm-up Questions

1) Write a balanced symbol equation for aerobic respiration.
2) Give three uses for the energy released in respiration.
3) Which cells in the body respire?
4) What is the name of the oxygen carrying substance in red blood cells?
5) Explain the difference between systolic and diastolic blood pressure.

Exam Questions

1   The table below describes the functions of different parts of the circulatory system.
     Match the words **A**, **B**, **C** and **D** with the numbers **1 - 4** in the table.

   A   Capillaries

   B   Heart

   C   Arteries

   D   Veins

Structure	Function
1	Pumps blood around the body
2	Return blood to the heart
3	Take blood away from the heart
4	Carry blood to every cell in the body

(4 marks)

2   Andrea is a sprinter. During a race Andrea's body can't supply enough oxygen to her
     muscles for aerobic respiration, so they start to respire anaerobically as well.

   (a)   Write a word equation for anaerobic respiration.

(1 mark)

   (b)   Anaerobic respiration is not as efficient as aerobic respiration.
         Explain why is it still useful to Andrea.

(1 mark)

   (c)   What problems could Andrea experience as a result of anaerobic respiration?

(1 mark)

   (d)   (i)   For a short time after the race, Andrea continues to breathe heavily.
               Explain why.

(2 marks)

         (ii) Her heart rate also remains high for a short time. Explain why.

(2 marks)

Digestion

Digestion is the breaking down of the nutrients in your food, so that they can be absorbed.

Big molecules are broken down into smaller ones

1) The aim of the game is to get all the <u>nutrients</u> from your food <u>into your blood</u>.

2) First the big lumps of food are <u>physically digested</u> so they can pass easily through the digestive system. This basically means <u>chewing</u> it in the mouth and <u>churning</u> it about in the stomach.

3) Then you use <u>chemical digestion</u> to break down molecules that are <u>too big</u> to pass through cell membranes.

4) This involves using <u>enzymes</u> — biological <u>catalysts</u> that break down the <u>big molecules</u> into <u>smaller ones</u>.

There are three main types of digestive enzyme

1) Carbohydrases

Carbohydrases break down <u>big carbohydrates</u> (e.g. starch) into <u>simple sugars</u>.

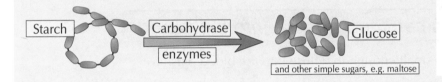

Starch → Carbohydrase enzymes → Glucose
and other simple sugars, e.g. maltose

They're present in <u>two</u> places:
1) The <u>mouth</u>
2) The <u>small intestine</u>

2) Proteases

Proteases convert <u>proteins</u> into <u>amino acids</u>.

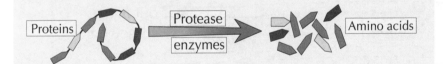

Proteins → Protease enzymes → Amino acids

They're present in <u>two</u> places:
1) The <u>stomach</u> (where it's called pepsin)
2) The <u>small intestine</u>

3) Lipases

Lipases convert <u>fats</u> into <u>fatty acids and glycerol</u>.

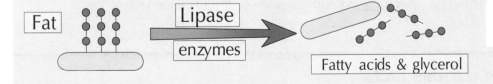

Fat → Lipase enzymes → Fatty acids & glycerol

They're present in the <u>small intestine</u>.

Digestion

Enzymes aren't the end of the story... There are a few more chemicals you need to know about, and then there's how molecules get from the gut to the rest of the body.

There are other **chemicals** in the body that **help enzymes work**

1) Stomach acid lowers the pH in the stomach, giving the right conditions for the digestive enzymes to work.

2) Bile is made in the liver and stored in the gall bladder.
It helps digestion in the small intestine in two ways:

- Bile is alkaline. It neutralises the acid from the stomach to make conditions right for the enzymes in the small intestine to work.

- It also emulsifies fats. In other words it breaks the fat into tiny droplets. This gives a much bigger surface area of fat for the lipase enzymes to work on.

The **small molecules** can then **diffuse** into the **blood**

1) Glucose and amino acids are small enough to diffuse into the blood plasma.

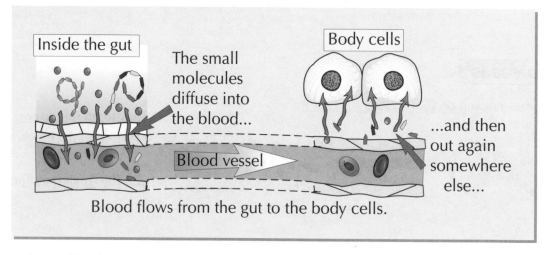

Blood flows from the gut to the body cells.

2) The products of fat digestion can't get into the blood plasma so they diffuse out of the gut (intestines) and into a fluid called lymph, in the lymphatic system. From here they're emptied into the blood.

3) The nutrients then travel to where they're needed, and then diffuse out again, e.g. glucose travels to muscles for respiration during exercise. It's all clever stuff.

Now get digesting those facts

Revision is a lot like digestion if you think about it — you break down the big topics into manageable chunks and then absorb the information. Imagine if you could eat this book and absorb all the facts in one sitting... it'd make life easier. But getting back to the real world... a mini-essay or two will help.

Warm-Up and Exam Questions

There's only one way to do well in the exam — learn the facts and then practise lots of exam questions to see what it'll be like on the big day. We couldn't have made that easier for you — so do it.

Warm-up Questions

1) Name one stage of physical digestion and state where it takes place.
2) Explain the role of enzymes in digestion.
3) What role does stomach acid play in digestion?
4) What are proteins broken down into?
5) How do the products of digestion pass across the gut wall?

Exam Questions

1 (a) Which of the following statements about bile is true?

    A It is made in the liver and stored in the pancreas.

    B It is made in the gall bladder and stored in the pancreas.

    C It is made in the liver and stored in the gall bladder.

    D It is made in the pancreas and stored in the gall bladder.

(1 mark)

(b) Describe one way in which bile aids digestion.

(1 mark)

2 (a) Which enzymes are responsible for digesting complex carbohydrates?

(1 mark)

(b) Name two places where these enzymes are found.

(2 marks)

3 Many foods contain fats.

(a) (i) Which enzymes are responsible for fat digestion?

(1 mark)

(ii) Where are these enzymes found?

(1 mark)

(iii) What are the two products of fat digestion?

(1 mark)

(b) The products of fat digestion do not enter the blood plasma immediately. Describe how the products of fat digestion are removed from the gut.

(1 mark)

Diet and Exercise

Why is it some people can eat loads and not put on weight, while others only have to look at a chocolate bar and they're a pound heavier...

*A **balanced diet** does a lot to keep you healthy*

1) For good health, your diet must provide the <u>energy</u> you need (but <u>not more</u>) — see the next page.

2) But that's not all. Because the different <u>food groups</u> have different uses in the body, you need to have the right <u>balance</u> of foods as well.

So you need: ...enough <u>carbohydrates</u> and <u>fats</u> to <u>keep warm</u> and provide <u>energy</u>,

...enough <u>protein</u> for <u>growth</u>, <u>cell repair</u> and <u>cell replacement</u>,

> You can calculate the <u>recommended daily allowance</u> (RDA) of <u>protein</u>:
>
> $$\text{RDA (g)} = 0.75 \times \text{body mass (kg)}$$
>
> E.g. calculate the recommended daily allowance of protein for someone who weighs 69 kg.
>
> $$\text{RDA} = 0.75 \times 69 = 51.75 \text{ g}$$

...enough <u>fibre</u> to keep everything moving <u>smoothly</u> through your digestive system,

...and tiny amounts of various <u>vitamins</u> and <u>minerals</u> to keep your skin, bones, blood and everything else generally healthy.

*People with **unbalanced diets** are said to be **malnourished***

1) Being <u>malnourished</u> is not the same as <u>starvation</u>.

2) Malnourished people can be <u>fat</u> or <u>thin</u>, or unhealthy in other ways.

> For example, a lack of <u>vitamin C</u> can cause scurvy, a <u>deficiency disease</u> which causes problems with the skin, joints and gums.

Citrus fruits are a good source of vitamin C.

3) Different <u>deficiency diseases</u> are caused by a lack of other nutrients.

You need to learn the components of a balanced diet

It's no good just eating the same thing all the time — to make sure you get everything your body needs, you should vary your diet. Carbohydrates, fats and proteins are fairly easy to come by, but a lot of people don't get enough fibre, vitamins and minerals. That's why dietary supplements are so popular (although the only way to get enough fibre is by eating lots of fruit, veg and cereals...)

Diet and Exercise

Not everybody has the same dietary requirements, it all depends on who you are and what you do.

People's energy needs vary because of **who they are**...

1) You need <u>energy</u> to fuel the chemical reactions in the body that keep you alive. These reactions are called your <u>metabolism</u>, and the speed at which they occur is your <u>metabolic rate</u>.

2) There are slight variations in the <u>resting metabolic rate</u> of different people. For example, <u>muscle</u> needs more energy than <u>fatty tissue</u>, which means (all other things being equal) people with a higher proportion of muscle to fat in their bodies will have a <u>higher</u> metabolic rate.

3) However, the <u>bigger</u> you are, the <u>more energy</u> your body needs to be supplied with — so bigger, heavier people normally have a <u>higher</u> metabolic rate.

4) <u>Men</u> tend to have a slightly <u>higher</u> rate than <u>women</u> — they're slightly <u>bigger</u> and have a larger proportion of <u>muscle</u>. Other <u>genetic factors</u> may also have some effect.

5) And regular <u>exercise</u> can boost your resting metabolic rate because it <u>builds muscle</u>.

...and because of **what they do**

1) When you <u>exercise</u>, you obviously need more <u>energy</u> — so your <u>metabolic rate</u> goes up during exercise and stays high for <u>some time</u> after you finish (particularly if the exercise is strenuous).

2) So people who have more <u>active</u> jobs need more <u>energy</u> on a daily basis — builders require more energy per day than office workers, for instance. The box below shows the average kilojoules burned per minute when doing different activities.

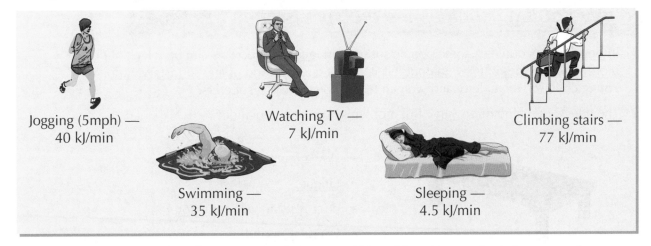

Jogging (5mph) — 40 kJ/min

Watching TV — 7 kJ/min

Climbing stairs — 77 kJ/min

Swimming — 35 kJ/min

Sleeping — 4.5 kJ/min

3) The <u>temperature</u> can also affect your metabolic rate. When it's <u>cold</u>, your body has to produce more heat (which requires energy) — this <u>increases</u> your metabolic rate.

4) All these factors have an effect on the amount of <u>energy</u> your <u>diet</u> should contain. If you do <u>little exercise</u> and it's <u>hot</u> outside, you're going to need <u>less energy</u> than if you're constantly on the go in a cold country.

Diet tip — the harder you revise the more calories you burn

Exercise is important as well as diet — people who <u>exercise regularly</u> are usually <u>fitter</u> than people who don't. But being <u>fit</u> isn't the same as being <u>healthy</u> — e.g. you can be fit as a fiddle and slim, but <u>malnourished</u> at the same time because your diet isn't <u>balanced</u>.

Diet Problems

Health problems due to the wrong kind of <u>diet</u> are different in different parts of the world. In some countries the problem is <u>too much</u> of the <u>wrong kind</u> of food, in others the problem is not having <u>enough</u>.

In **developed** countries the problem is **too much food**

1) In <u>developed</u> countries, obesity is becoming a serious problem. In the UK, 1 in 5 adults are obese, with obesity contributing to the deaths of over 30 000 people each year in England alone.

2) Hormonal problems can lead to obesity, though the usual cause is a <u>bad diet</u>, <u>overeating</u> and a <u>lack</u> of <u>exercise</u>.

3) Health problems that can arise as a result of obesity include:

- <u>arthritis</u> (inflammation of the joints)
- <u>diabetes</u> (inability to control blood sugar levels)
- <u>high blood pressure</u>
- <u>heart disease</u>
- It's also a risk factor for some kinds of <u>cancer</u>.

4) The National Health Service spends loads each year treating <u>obesity-related</u> conditions. And more is lost to the <u>economy</u> generally due to absence from work.

In **developing** countries the problem is often **too little**

1) In <u>developing</u> countries, some people suffer from <u>lack</u> of food. This can be a lack of one or more <u>specific types</u> of food (<u>malnutrition</u>), or not enough food <u>of any sort</u> (<u>starvation</u>). Young children, the elderly and women tend to be the worst sufferers.

2) The effects of malnutrition <u>vary</u>. But problems commonly include:

- <u>slow growth</u> (in children)
- <u>fatigue</u>
- poor <u>resistance</u> to <u>infection</u>
- <u>irregular periods</u> in women

3) Eating <u>too little protein</u> can cause a condition called <u>kwashiorkor</u>.

Diets in many parts of the world are deficient in protein, especially in <u>poorer developing countries</u>, as protein-rich foods are often <u>too expensive</u> to buy.

There's enough food around, it's just distributing it properly

Total world food production is enough to feed everyone on the planet, but some countries end up with much more food than others. It's a complicated problem involving things like climate, education, conflict and money, but it means that around 850 million people worldwide are undernourished.

Diet Problems

Having too much or too little of the wrong kinds of foods can have some serious health impacts.

Body mass index *indicates if you're* **under-** *or* **overweight**

Body mass index (BMI) is calculated from a person's height and weight.

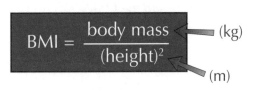

$$BMI = \frac{body\ mass}{(height)^2}$$

(kg)

(m)

Body Mass Index	Weight Description
below 18.5	underweight
18.5 - 24.9	normal
25 - 29.9	overweight
30 - 40	moderately obese
above 40	severely obese

The table shows how BMI is used to classify people's weight. →

Example

Calculate the BMI of someone who is 1.7 m tall and weighs 76 kg.

$76 \div (1.7)^2 = 26.3$. So using the table this person is overweight.

BMI isn't always reliable — athletes can have lots of muscle, which weighs more than fat, so they can come out with a high BMI even though they're not overweight.

High cholesterol *and* salt *levels are* risk factors *for* heart disease

1) Cholesterol is a fatty substance that's essential for good health. It's found in every cell in the body.

2) But a high cholesterol level in the blood causes an increased risk of various problems — like coronary heart disease. This is due to blood vessels getting clogged with fatty cholesterol deposits.

3) The liver is really important in controlling the amount of cholesterol in the body. It makes new cholesterol, and removes it from the blood so that it can be eliminated from the body.

The amount the liver makes depends on your diet and inherited factors.

4) Eating too much salt may cause high blood pressure for about 30% of the UK population. However, it's not always easy to keep track of how much salt you eat — most of the salt you eat is probably in processed foods. The salt you sprinkle on your food makes up quite a small proportion.

5) And just to complicate things, on food labels, salt is usually listed as sodium.

Lay off the salty snacks when you're practicing the BMI equation

It's very difficult to accurately measure how under- or overweight someone is, simply because everyone's different — there's no real rule about your 'perfect' weight because we all have different bone structures, muscle:fat ratios etc. BMI is useful though, because it provides a rough measure — if you're in the low 20s, you're probably alright, but that's about as accurate as it gets.

Warm-Up and Exam Questions

By the time the big day comes you need all the facts in these warm-up questions and all the exam questions like the back of your hand. It's not a barrel of laughs, but it's the only way to get good marks.

Warm-up Questions

1) What are the most common causes of obesity?
2) Why is it important to eat enough fibre as part of a balanced diet?
3) Which organ is responsible for controlling the amount of cholesterol in the body?
4) Why does high blood cholesterol increase the risk of heart disease?
5) What is the difference between malnutrition and starvation?

Exam Questions

1   Which of the following statements about metabolic rate is **not** true?

   A   People who weigh more have a higher metabolic rate.

   B   In general, women have a higher metabolic rate than men.

   C   A person's metabolic rate is affected by the temperature of their surroundings.

   D   Regular exercise increases resting metabolic rate.

(1 mark)

2   (a)   (i)   James is 185 cm tall and weighs 81 kg. Calculate his BMI.
            BMI = body mass (kg) $\div$ height^2 (m)

(1 mark)

          (ii) Use the table to give his weight description.

(1 mark)

BMI	Weight description
below 18.5	underweight
18.5 - 24.9	normal
25.0 - 29.9	overweight
30.0 - 40.0	obese
over 40.0	severely obese

   (b)   Obesity increases the risk of various health problems. Name one such problem.

(1 mark)

3   Kwashiorkor is a condition caused by too little protein in the diet.

   (a)   Suggest why Kwashiorkor is more common in developing countries.

(1 mark)

   (b)   Name three things protein is needed for in the body.

(3 marks)

   (c)   The recommended daily allowance (RDA) of protein can be calculated from the
         formula: RDA protein (g) = 0.75 x body mass (kg).

         What is the RDA of protein for a person who weighs 75 kg?

(1 mark)

Health Claims

It's sometimes hard to figure out if <u>health claims</u> or <u>adverts</u> are <u>true</u> or not.

New day, new food claim — *it can't all be true*

1) To get you to buy a product, advertisers aren't allowed to make claims that are <u>untrue</u> — that's <u>illegal</u>.

2) But they do sometimes make claims that could be <u>misleading</u> or difficult to <u>prove</u> (or <u>disprove</u>).

   For example, some claims are just <u>vague</u> (calling a product "light" for instance — does that mean low calorie, low fat, something else...).

   Alternatively, they might call a breakfast cereal "low fat", and that'd be <u>true</u>. But that could suggest that <u>other</u> breakfast cereals are high in fat — when in fact they're not.

3) And every day there's a new <u>food scare</u> in the papers (eeek — we're all doomed). Or a new <u>miracle food</u> (phew — we're all saved).

4) It's not easy to decide what to <u>believe</u> and what to <u>ignore</u>. But these things are worth looking for:

   > a) Is the report a <u>scientific study</u>, published in a <u>reputable journal</u>?
   >
   > b) Was it written by a <u>qualified person</u> (not connected with the food producers)?
   >
   > c) Was the <u>sample</u> of people asked/tested <u>large enough</u> to give reliable results?
   >
   > d) Have there been <u>other studies</u> which found <u>similar results</u>?

5) A "yes" to one or more of these is a good sign.

Not all diets are *scientifically proven*

With each new day comes a new celebrity-endorsed diet. It's a wonder anyone's overweight.

1) A common way to promote a new <u>diet</u> is to say, "Celebrity A has lost x pounds using it".

2) But effectiveness in <u>one person</u> doesn't mean much. Only a <u>large survey</u> can tell if a diet is more or less effective than just <u>eating less</u> and <u>exercising more</u> — and these aren't done often.

Example

The <u>Atkins diet</u> was high profile, and controversial — so it got investigated. People on the diet certainly lost weight. But the diet's effect on general health (especially <u>long-term</u> health) has been questioned. The jury's still out.

3) Weight loss is a <u>complex</u> process. But just like with food claims, the best thing to do is look at the evidence in a scientific way.

Health Claims

The same rules apply when looking into claims about drugs — look at all the evidence in a scientific way.

It's the same when you look at claims about **drugs**

Claims about the effects of <u>drugs</u> (both medical and illegal ones) also need to be looked at <u>critically</u>. But at least here the evidence is usually based on <u>scientific</u> research.

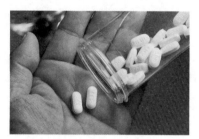

Statins

1) There's evidence that drugs called <u>statins</u> lower <u>blood cholesterol</u> and significantly lower the risk of <u>heart disease</u> in diabetic patients.

2) The original research was done by <u>government</u> scientists with <u>no connection</u> to the manufacturers. And the <u>sample</u> was <u>big</u> — 6000 patients.

So control groups were used. And the results were reproducible.

3) It compared <u>two groups</u> of patients — those who <u>had</u> taken statins and those who <u>hadn't</u>. Other studies have since <u>backed up</u> these findings.

But research findings are not always so clear cut...

Cannabis

1) Many scientists have looked at whether <u>cannabis</u> use causes brain damage and mental health problems or leads to further drug taking. The results <u>vary</u>, and are sometimes open to different <u>interpretations</u>.

2) Basically, until more definite scientific evidence is found, no one's <u>sure</u>.

'Brad Pitt says it's great' is NOT scientific proof

Learn what to look out for before you put too much faith in what you read. Then buy my book — 100% of the people I surveyed (i.e. both of them) said it had no negative effect <u>whatsoever</u> on their overall wellbeing! Buy now and receive a free sample of my patented <u>brain-stimulating</u> revision drink.

Drugs

Drugs alter what goes on in your body. Your body's essentially a seething mass of chemical reactions — drugs can interfere with these reactions, sometimes for the better, sometimes not.

Drugs can be **beneficial** or **harmful**, **legal** or **illegal**

1) Drugs are substances which alter the chemical reactions in the body. Some drugs are medically useful, such as antibiotics (e.g. penicillin).

2) But many drugs are dangerous if misused (this goes for both illegal drugs and legal ones). That could mean problems with either your physical or mental health.

3) This is why you can buy some drugs over the counter at a pharmacy, others are restricted so you can only get them on prescription (your doctor decides if you should have them), and others are illegal.

4) Some people get addicted to some drugs — this means they have a physical need for that drug, and if they don't get it they experience withdrawal symptoms.

5) It's not just illegal drugs that are addictive — many legal ones are too, e.g. caffeine. Caffeine withdrawal symptoms include irritability and shaky hands.

6) Tolerance develops with some drugs — the body gets used to having it and so you need a higher dose to give the same effect. This can happen with legal drugs (e.g. alcohol), and illegal drugs (e.g. heroin).

Illegal drugs are classified into **three** main categories

1) Some drugs are illegal — usually because they're considered to be dangerous.

2) In the UK, they either belong to Class A, B or C. Which class a drug is in depends on how dangerous it is thought to be — Class A drugs are the most dangerous.

Class A drugs

   heroin, LSD, ecstasy and cocaine

Class B drugs

   amphetamines (speed)

   (Amphetamines are class A if prepared for injection.)

Class C drugs

   cannabis, anabolic steroids and tranquillisers

This isn't too difficult — it's really just learning your ABCs

There's been a lot of debate recently about how useful the illegal drug classification system is. A lot of people argue that it isn't based on scientific evidence, and that it should be redesigned. Luckily for you, that's not going to be in the exams — just learn the current system and you'll be fine.

Drugs

So far you've looked at how drugs can be <u>classified</u> depending on how <u>dangerous</u> they're thought to be, but this page is all about what they actually <u>do</u>.

Drugs can affect your **behaviour**

1) A lot of drugs affect your <u>nervous system</u>. Drugs can interfere with the way <u>signals</u> are sent around your body from <u>receptors</u> to the <u>brain</u>, and from the <u>brain</u> to <u>muscles</u> (see page 6).

2) The effects of drugs on the nervous system can alter <u>behaviour</u> (which has the potential to cause <u>danger</u> — either for the person who took the drug, or for others).

3) For example, <u>driving</u> and operating <u>machinery</u> aren't safe if you've taken certain drugs — e.g. alcohol, tranquillisers or cannabis (see page 43 for more info about alcohol).

4) Some drugs (e.g. alcohol) can also affect people's <u>judgement</u>. This could mean someone just '<u>losing their inhibitions</u>' — relaxing a bit at a party, for instance.

5) But it could mean they take more <u>risks</u> — e.g. <u>sharing needles</u> and having <u>unprotected sex</u> are more likely to happen under the influence of drink or drugs, increasing the risk of viral infections like <u>HIV</u>.

6) Drug abuse can also affect your <u>immune system</u> — making infections more <u>likely</u>.

Sedatives slow you down

These are also called depressants.

— e.g. alcohol, barbiturates, solvents, temazepam.

These <u>decrease</u> the <u>activity of the brain</u>. This slows down the <u>responses</u> of the <u>nervous system</u>, causing <u>slow reactions</u> and <u>poor judgement</u> of speed and distances.

Stimulants speed you up

— e.g. <u>nicotine</u>, <u>ecstasy</u>, <u>caffeine</u>.

These do the opposite of depressants — they <u>increase</u> the <u>activity of the brain</u>. This makes you feel more <u>alert</u> and <u>awake</u>.

Learn all the grim facts

Many people regularly have <u>sedatives</u> and <u>stimulants</u>, the most common ones being alcohol and caffeine. Those two don't seem to be too bad for you in small doses — it's <u>misuse</u> that can get you into trouble (e.g. having 25 coffees and 17 pints a day is liable to put you in hospital).

Medical Drugs

Drugs have medical uses too, obviously. But before they can be used, they have to be tested...

Medical drugs have to be thoroughly tested

New drugs are constantly being developed. But before they can be given to the general public, they have to go through a thorough testing procedure. This is what usually happens...

Computer models are often used in the early stages — these simulate a human's response to a drug.

This can identify promising drugs to be tested in the next stage (but sometimes it's not as accurate as actually seeing the effect on a live organism).

Drugs are then developed further by testing on human tissues in the lab.

However, you can't use human tissue to test drugs that affect whole or multiple body systems, e.g. testing a drug for blood pressure must be done on a whole animal because it has an intact circulatory system.

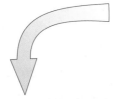

The next step is to develop and test the drug using live animals. The law in Britain states that any new drug must be tested on two different live mammals.

Some people think it's cruel to test on animals, but others believe this is the safest way to make sure a drug isn't dangerous before it's given to humans.

But some people think that animals are so different from humans that testing on animals is pointless.

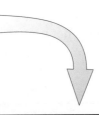

After the drug has been tested on animals it's tested on human volunteers in a clinical trial — this should determine whether there are any side effects.

There's more about clinical trials on the next page.

Try to be as open minded as you can with this one

Most people have an opinion of some kind when it comes to testing drugs on animals. Whatever yours is, you need to be able to give both sides of the story. So make sure you learn how animal testing fits in with the other tests new drugs are put through — learn the whole page.

Medical Drugs

So, new drugs have to go through fairly <u>thorough</u> testing before they get anywhere near <u>humans</u>. And even when they do get to the clinical trial stage, you can't always test <u>everything</u>.

Clinical trials involve two groups of patients

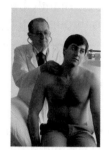

1) One group is given the <u>new drug</u>, the other is given a <u>placebo</u> (a 'dummy treatment' that looks like the real drug but doesn't do anything). This is done so scientists can see the actual difference the drug makes — it allows for the <u>placebo effect</u> (when the patient expects the treatment to work and so <u>feels better</u>, even though the treatment isn't doing anything).

2) Clinical trials are <u>blind</u> — the patient in the study <u>doesn't know</u> whether they're getting the drug or the placebo. In fact, they're often <u>double blind</u> — neither the <u>patient</u> nor the <u>scientist</u> knows until all the results have been gathered.

Things have gone wrong in the past

An example of what can happen when drugs are not thoroughly tested is the case of <u>thalidomide</u> — a drug developed in the 1950s.

1) Thalidomide was intended as a <u>sleeping pill</u>, and was tested for that use. But later it was also found to be effective in relieving <u>morning sickness</u> in pregnant women.

2) Unfortunately, thalidomide <u>hadn't</u> been <u>tested</u> for this use, and so it wasn't known that it could pass through the placenta and affect the <u>foetus</u>, causing <u>stunted growth</u> of the foetus's arms and legs. In some cases, babies were born with no arms or legs at all.

3) About <u>10 000</u> babies were affected by thalidomide, and only about <u>half</u> of them survived.

4) The drug was <u>banned</u>, and more <u>rigorous</u> testing procedures were introduced.

5) Thalidomide has recently been reintroduced — as a treatment for <u>leprosy</u>, <u>AIDS</u> and certain <u>cancers</u>. But it can't be used on pregnant women.

Developing new drugs is expensive

1) New drugs are often very sophisticated, and it can take <u>many years</u> to <u>develop</u> and <u>test</u> a drug to the stage where it can be put into use.

2) Also, most potential drugs are <u>rejected</u> during the trials.

3) All of this adds to the <u>cost</u> of coming up with a drug that can be used <u>safely</u> on humans.

A little learning is a dangerous thing

Thalidomide was an attempt to <u>improve</u> people's lives which then caused some pretty tragic knock-on effects. Could the same thing happen <u>today</u>? Well, maybe not the exact same thing, but there's no such thing as <u>perfect</u> knowledge — you can never eliminate risk <u>completely</u>.

Warm-Up and Exam Questions

There's no point in skimming through the section and glancing over the questions. Do the warm-up questions and go back over any bits you don't know. Then practise and practise the exam questions.

Warm-up Questions

1) What is a drug?
2) Give four things that should be considered when deciding if a health claim is reliable.
3) What is a placebo?
4) Name one legal drug to which people may become addicted.
5) Why is developing new drugs an expensive process?

Exam Questions

1 (a) Which of these drugs is a sedative?

    A nicotine

    B alcohol

    C caffeine

    D ecstasy

(1 mark)

  (b) What effect do sedatives have on the nervous system?

(1 mark)

2 (a) Give four stages of testing that a new drug will usually go through before it can be sold to the general public.

(4 marks)

  (b) Describe how a typical double blind clinical trial works.

(2 marks)

3 (a) It is possible to build up a tolerance and/or become addicted to both illegal and legal drugs. Explain what is meant by:

    (i) addiction

(1 mark)

    (ii) tolerance

(1 mark)

  (b) Illegal drugs in the UK are placed into three main categories.

    (i) Which class of drugs is the most dangerous?

(1 mark)

    (ii) Give an example of a drug in this class.

(1 mark)

  (c) Why are people who abuse drugs at greater risk of picking up infections?

(1 mark)

Smoking and Alcohol

Drugs are also used <u>recreationally</u>. Some of these are legal, others illegal. And some are more <u>harmful</u> than others. But two drugs that have a massive impact on people and society are both <u>legal</u>.

*Smoking **tobacco** can cause quite a few **problems***

1) Tobacco smoke contains <u>carbon monoxide</u> — this <u>combines</u> irreversibly with <u>haemoglobin</u> in blood cells, meaning the blood can carry <u>less oxygen</u>. In pregnant women, this can deprive the <u>foetus</u> of oxygen, leading to the baby being born <u>underweight</u>.

2) Tobacco smoke also contains carcinogens — chemicals that can lead to <u>cancer</u>. Lung cancer is way more common among smokers than nonsmokers.

3) Disturbingly, the <u>incidence rate</u> (the number of people who get lung cancer) and the <u>mortality rate</u> (the number who die from it) aren't massively different — lung cancer kills <u>most</u> of the people who get it.

4) Smoking also causes <u>disease</u> of the <u>heart</u> and <u>blood vessels</u> (leading to <u>heart attacks</u> and <u>strokes</u>), and damage to the <u>lungs</u> (leading to diseases like <u>emphysema</u> and <u>bronchitis</u>).

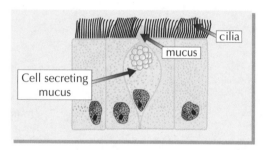

5) And the <u>tar</u> in cigarettes damages the <u>cilia</u> (little hairs) in your lungs and windpipe (see p48). These hairs, along with <u>mucus</u>, catch a load of <u>dust</u> and <u>bacteria</u> before they reach the lungs. When these cilia are damaged, <u>chest infections</u> are more likely.

6) And to top it all off, smoking tobacco is <u>addictive</u> — due to the <u>nicotine</u> in tobacco smoke.

*Smoking and lung cancer are now **known** to be linked*

1) In the first half of the 20th century it was noticed that <u>lung cancer</u> and the popularity of <u>smoking</u> increased <u>together</u>. And studies found that far more <u>smokers</u> than <u>nonsmokers</u> got lung cancer.

2) But it was just a <u>statistical correlation</u> at that time (see p285) — it didn't <u>prove</u> that smoking <u>caused</u> lung cancer. Some people (especially in the tobacco industry) argued that there was some <u>other</u> factor (e.g. a person's <u>genes</u>) which both caused lung cancer, and also made people more likely to smoke.

3) Later research eventually <u>disproved</u> these claims. Now even the tobacco industry has had to admit that smoking does <u>increase</u> the <u>risk</u> of lung cancer.

If that wasn't enough, it also makes you look cool

This page might make you think 'yeah yeah yeah heard it before', but there's no way you can make an <u>informed decision</u> about drug use without being informed. Whatever your take on this kind of thing, you need to know about it for your <u>exam</u>. So make sure you learn all the details.

Smoking and Alcohol

Many people see drinking alcohol as more acceptable than smoking tobacco, but excessive drinking seems to be on the increase, and so are drink-related crimes and injuries.

Drinking alcohol can do its share of damage too

1) The main effect of alcohol is to <u>reduce the activity</u> of the <u>nervous system</u> — slowing your reactions. It can also make you feel <u>less inhibited</u> — which can help people to socialise and relax with each other.

2) However, too much leads to <u>impaired judgement</u>, <u>poor balance</u> and <u>coordination</u>, <u>lack of self-control</u>, <u>unconsciousness</u> and even <u>coma</u>.

Alcohol in excess also causes <u>dehydration</u>, which can damage <u>brain cells</u>, causing a noticeable <u>drop</u> in <u>brain function</u>. And too much drinking causes <u>severe damage</u> to the <u>liver</u>, leading to <u>liver disease</u>.

3) There are <u>social</u> costs too. Alcohol is linked with loads of murders, stabbings and domestic assaults.

These two legal drugs have a massive impact

<u>Alcohol</u> and <u>tobacco</u> have a bigger impact in the UK than illegal drugs, as <u>so many</u> people take them.

1) Tobacco

The National Health Service spends loads on treating people with <u>lung diseases</u> caused by <u>smoking</u> (or passive smoking). Add to this the cost to businesses of people missing days from work, and the figures get pretty scary.

2) Alcohol

The same goes for <u>alcohol</u>. The costs to the NHS are huge, but are pretty small compared to the costs related to <u>crime</u> (police time, damage to people/property) and the <u>economy</u> (lost working days etc.).

Learn all this stuff — not just for the exam

So it's legal drugs that have the most impact on the country as a <u>whole</u> — when you take everything into consideration. Should the <u>Government</u> do more to reduce the number of people who smoke — or is it up to individual <u>people</u> what they do with their lives... there's no easy answer to that one.

Solvents and Painkillers

Two other groups of drugs you need to know about are solvents and painkillers. There's not too much you need to know, just remember a few examples of each and what both the groups do.

Solvents affect the lungs and neurones

1) Solvents are found in lighter fuel, spray paints, aerosols, thinners and dry cleaning fluids. They're useful chemicals, but can be misused as <u>drugs</u> (by <u>inhaling</u> the fumes).

2) Solvents act on the <u>nervous system</u>. Like alcohol, they're <u>depressants</u> — they slow down messages as they're passed along <u>neurones</u> (and can cause all sorts of other damage as well).

3) Solvent abuse often causes <u>brain damage</u> in the long term — this could show up as a personality change, sleeplessness or short-term memory loss, for example.

4) Most solvents also irritate the <u>lungs</u> and the <u>breathing passages</u>.

Paracetamol is a painkiller

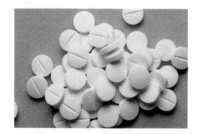

1) <u>Paracetamol</u> is a medicine that can <u>relieve</u> mild to moderate <u>pain</u>, and reduce <u>fever</u>.

2) Paracetamol is generally pretty <u>safe</u>, but an <u>overdose</u> can be <u>deadly</u>. <u>Paracetamol overdose</u> causes horrendous liver damage. If it isn't treated quickly (and I mean really quickly) it's <u>very</u> dangerous. And paracetamol's especially dangerous after alcohol, so it's not a good idea for hangovers.

3) A paracetamol overdose is <u>particularly dangerous</u> because the damage sometimes isn't apparent for <u>4-6 days</u> after the drug's been taken. By that time, it's <u>too late</u> — there's nothing doctors can do to repair the damage. Dying from liver failure takes several days, and involves <u>heavy-duty pain</u>.

4) Paracetamol in <u>normal doses</u> won't damage the liver (though <u>accidental</u> overdoses are quite common).

Drugs can kill you or cure you (or anything in between)

Paracetamol's a drug that most people come across fairly regularly, and as long as you read the label and make sure you don't take too many, it's perfectly safe. The same can be said for a lot of drugs, but when it comes to solvents, you really are better off just steering well clear (assuming you value your lungs and/or brain). Unless of course you're doing some spray painting, which is fine. As long as you don't spend all day inhaling paint fumes, or paint your eyes shut or something.

Solvents and Painkillers

There's a good chance you'll know most of things on this page already, but read through it thoroughly anyway to get an idea of what you need for the exams.

Opiates and *cannabinoids* are used as *painkillers*

Some types of painkillers can only be used under <u>medical supervision</u>.

Opiates

- Opiates include <u>opium</u>, <u>morphine</u> and <u>codeine</u>. They're all found in the opium poppy.
- Opiates are all <u>painkillers</u>. <u>Morphine</u>'s used by doctors — it's very effective. But just like heroin, morphine's very <u>addictive</u>, and so it's <u>illegal</u> without a prescription.

Cannabis

- Cannabis has been used as a medicine for centuries, but it's now <u>illegal</u>.
- For years, no one really <u>knew</u> what cannabis did inside the body — this was because research on <u>cannabinoids</u> (the active ingredients in cannabis) was tricky (due to <u>legal restrictions</u>).
- The situation changed when scientists discovered <u>receptors</u> in the body for cannabinoids. Recent research seems to suggest that cannabinoids do provide <u>benefit</u> for <u>some</u> patients (though for <u>most</u> people, there's probably something <u>better</u> available).

Different painkillers work in *different ways*

1) <u>Aspirin</u> and <u>ibuprofen</u> work by inhibiting the formation of <u>prostaglandins</u> (chemicals which cause <u>swelling</u>, and sensitise the endings of nerves that register pain).

2) <u>Paracetamol</u> seems to work in a <u>similar way</u> to aspirin and ibuprofen, but scientists <u>aren't really sure</u>.

3) <u>Opiates</u>, like morphine and codeine, are <u>very strong</u> painkillers. They work by interfering with the <u>mechanism</u> by which 'pain-sensing' nerve cells transmit messages. They also act on the <u>brain</u> to stop it sensing the pain.

You can learn this — take the pain, take the pain

Isn't it amazing that we're still not sure how paracetamol works... Apparently the pain-reducing effects of paracetamol were just discovered by accident. That's science for you — a series of accidents which add together to make amazing discoveries. Learn all the stuff, test yourself, and learn it again if need be.

Warm-Up and Exam Questions

Without a good warm-up you're likely to strain a brain cell or two. So take the time to run through these simple questions and get the basic facts straight before plunging into the exam questions.

Warm-up Questions

1) Name two organs that can be damaged by solvent abuse.
2) To which class of painkillers does morphine belong?
3) Name two lung diseases, other than cancer, caused by smoking.
4) What are carcinogens?
5) Suggest why there has been so little research into the medicinal use of cannabinoids?

Exam Questions

1   The following chemicals are all found in cigarette smoke.
     Explain what effect each has on the body.

   (a) Tar

(1 mark)

   (b) Carbon monoxide

(1 mark)

   (c) Nicotine

(1 mark)

2   (a) Describe how the following painkillers work on the body:

       (i)   Aspirin

(1 mark)

       (ii) Morphine

(1 mark)

   (b) Explain why is it important not to take more than the recommended daily
        dose of paracetamol?

(1 mark)

3   In the UK it is illegal to drive if your blood alcohol concentration exceeds 80 mg of
     alcohol per 100 ml of blood.

   (a) Explain the effect alcohol has on the body, that increases the risk of having a car
        accident when drink driving?

(2 marks)

   (b) Give two long term health effects of excessive alcohol consumption.

(2 marks)

   (c) Other than drink-related driving accidents, give two ways in which excessive
        alcohol consumption has a negative effect on society.

(2 marks)

Causes of Disease

An <u>infectious</u> disease is a disease that can be <u>transmitted</u> from one person to another — either <u>directly</u> (person to person), or <u>indirectly</u> (where some kind of <u>carrier</u> is involved, e.g. mosquitoes spread malaria, and certain bacteria are passed on in food or water).

Infectious diseases are caused by *pathogens*

1) <u>Pathogens</u> are <u>microorganisms</u> (<u>microbes</u>) that cause <u>disease</u>.

2) They include some <u>bacteria</u>, <u>protozoa</u> (certain single-celled creatures), <u>fungi</u> and <u>viruses</u>.

3) All pathogens are <u>parasites</u> — they live off their host and give nothing in return.

4) Microorganisms can <u>reproduce very fast</u> inside a host organism.

Bacteria and viruses are **very different**

...but they can both multiply quickly inside your body — they love the warm conditions.

Bacteria are very small **living cells**

1) Bacteria are <u>very small cells</u> (about 1/100th the size of your body cells), which can reproduce rapidly inside your body.

2) They make you <u>feel ill</u> by doing <u>two</u> things — <u>damaging your cells</u> and <u>producing toxins</u> (poisons).

3) But... some bacteria are <u>useful</u> if they're in the <u>right place</u>, like in your digestive system.

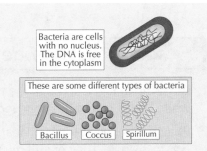

Bacteria are cells with no nucleus. The DNA is free in the cytoplasm

These are some different types of bacteria

Bacillus | Coccus | Spirillum

Viruses are **not cells** *— they're much* **smaller**

1) Viruses are <u>not cells</u>. They're <u>tiny</u>, about 1/100th the size of a bacterium. They're usually no more than a <u>coat of protein</u> around some <u>genetic material</u>.

2) They <u>replicate themselves</u> by invading a cell and using the DNA it contains to produce many <u>copies</u> of themselves. The cell will usually then <u>burst</u>, releasing all the new viruses.

3) This <u>cell damage</u> is what makes you feel ill.

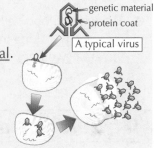

genetic material
protein coat
A typical virus

Other **health disorders** *can be caused in* **various ways**

1) <u>Vitamin deficiency</u>, e.g. you can get <u>scurvy</u> if you don't get enough <u>vitamin C</u>.

2) <u>Mineral deficiency</u>, e.g. a lack of <u>iron</u> in the diet can lead to <u>anaemia</u>. Iron is needed to make the protein <u>haemoglobin</u> (which carries <u>oxygen</u> in the red blood cells).

3) <u>Genetic inheritance</u> of disorders (see p66), e.g. <u>red-green colour blindness</u> (sufferers find it hard to distinguish between red and green) and <u>haemophilia</u> (a blood clotting disorder).

4) <u>Body disorders</u> are caused by body cells not working properly, e.g. <u>diabetes</u> (see p21) and <u>cancer</u>.

> Cancer is caused by body cells growing <u>out of control</u>. This forms a <u>tumour</u> (a mass of cells). Tumours can either be <u>benign</u> or <u>malignant</u>:
> 1) Benign — This is where the tumour grows until there's no more room. The cells <u>stay</u> where they are. This type <u>isn't</u> normally dangerous.
>
> 2) Malignant — This is where the tumour grows and can <u>spread</u> to other sites in the body. Malignant tumours are <u>dangerous</u> and can be fatal.

The Body's Defence Systems

Your body is constantly fighting off attack from all sorts of nasties — yep, things really are out to get you. The body has <u>three</u> lines of defence to stop things causing disease.

The *first line of defence* stops pathogens *entering* the body

The first line of defence consists mostly of <u>physical barriers</u> — they stop <u>foreign bodies</u> getting in.

1) The Skin
<u>Undamaged skin</u> is a very effective barrier against microorganisms. And if it gets <u>damaged</u>, blood <u>clots</u> quickly to <u>seal cuts</u> and keep microorganisms <u>out</u>.

3) The Respiratory System
The nasal passage and trachea are lined with <u>mucus</u> and <u>cilia</u> which catch <u>dust</u> and <u>bacteria</u> before they reach the lungs.

2) The Eyes
<u>Eyes</u> produce (in <u>tears</u>) a chemical called <u>lysozyme</u> which <u>kills bacteria</u> on the surface of the eye. This is a <u>chemical barrier</u> — not a physical one.

The *second line of defence* is *non-specific white blood cells*

1) Anything that gets through the first line of defence and into the body should be picked up by white blood cells called <u>phagocytes</u> (a <u>chemical</u> barrier).

2) Phagocytes detect things that are '<u>foreign</u>' to the body, e.g. microbes. They <u>engulf microbes</u> and <u>digest them</u>.

3) Phagocytes are <u>non-specific</u> — they attack anything that's not meant to be there.

White Blood Cell

microbes

4) The white blood cells also trigger an <u>inflammatory response</u>. <u>Blood flow</u> to the infected area is <u>increased</u> (making the area <u>red</u> and <u>hot</u>), and <u>plasma</u> leaks into the damaged tissue (which makes the area <u>swell up</u>) — this is all so that the right cells can get to the area to <u>fight</u> the infection.

The *third line of defence* is *specific white blood cells*

Some produce *antibodies*

1) Every invading cell has unique molecules (called <u>antigens</u>) on its surface.

2) When certain white blood cells come across a <u>foreign antigen</u> (i.e. one it doesn't recognise), they will start to produce <u>proteins</u> called <u>antibodies</u> to lock on to the invading cells and mark them out for destruction by other white blood cells. The antibodies produced are specific to that type of antigen — they won't lock on to any others.

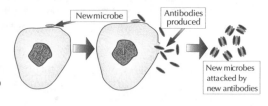

New microbe

Antibodies produced

New microbes attacked by new antibodies

3) Antibodies are then produced <u>rapidly</u> and flow round the body to mark all similar bacteria or viruses.

4) Some of these white blood cells stay around in the blood after the original infection has been fought. They can reproduce very fast if the <u>same</u> antigen enters the body for a <u>second</u> time. That's why you're immune to <u>most</u> diseases if you've already had them — the body carries a "<u>memory</u>" of what the antigen was like, and can quickly produce loads of antibodies if you get infected again.

Some produce *antitoxins*
These counter toxins produced by <u>invading microbes</u>.

Immunisation

Immunisation changed the way we deal with disease. Not bad for a little jab.

Immunisation — protects from future infections

1) When you're infected with a new <u>microorganism</u>, it takes your white blood cells a few days to <u>learn</u> how to deal with it. But by that time, you can be pretty <u>ill</u>.

2) <u>Immunisation</u> involves injecting <u>dead</u> or <u>inactive</u> microorganisms. These carry <u>antigens</u>, which cause your body to produce <u>antibodies</u> to attack them — even though the microorganism is <u>harmless</u> (since it's dead or inactive). For example, the MMR vaccine contains <u>weakened</u> versions of the viruses that cause <u>measles</u>, <u>mumps</u> and <u>rubella</u> (German measles) stuck together.

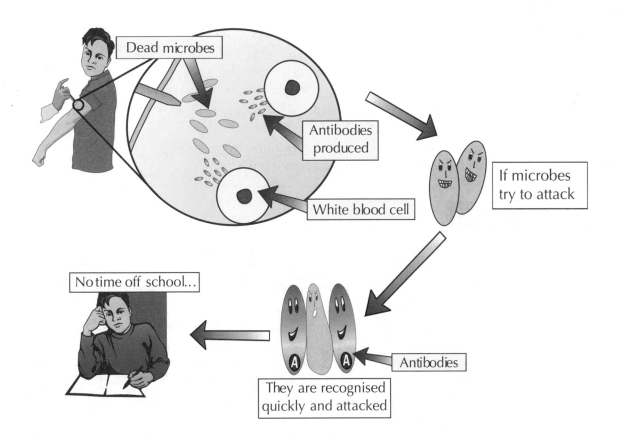

Dead microbes

Antibodies produced

White blood cell

If microbes try to attack

Antibodies

They are recognised quickly and attacked

No time off school...

3) But if live microorganisms of the same type appear after that, the white blood cells can <u>rapidly</u> mass-produce antibodies to help kill off the pathogen.

4) Vaccinations "wear off" over time. So <u>booster</u> injections can be given to increase levels of antibodies again.

Immunisation

Immunisation provides huge benefits to society.
Having said that, there are a few risks that you should be aware of.

Immunisation is classed as *active immunity*

1) <u>Active immunity</u> is where the immune system makes <u>its own antibodies</u> after being stimulated by a <u>pathogen</u>. It includes becoming <u>naturally immune</u> (see page 48) and <u>artificially immune</u> (immunisation). Active immunity is usually <u>permanent</u>.

2) <u>Passive immunity</u> is where you use antibodies <u>made by another organism</u>, e.g. antibodies are passed from mother to baby through <u>breast milk</u>. Passive immunity is only <u>temporary</u>.

There are *benefits* and *risks* associated with *immunisation*

Benefits

1) Immunisation <u>stops you from getting ill</u>... a pretty obvious benefit.

2) Vaccinations mean we don't have to deal with a problem once it's happened — we can <u>prevent</u> it happening in the first place. Vaccines have helped <u>control</u> lots of infectious diseases that were once <u>common</u> in the UK (e.g. polio, measles, whooping cough, rubella, mumps, tetanus, TB...).

3) If an outbreak does occur, vaccines can <u>slow down</u> or <u>stop</u> the spread (if people don't catch the disease, they won't pass it on).

Risks

1) There can be <u>short-term side effects</u>, e.g. <u>swelling</u> and <u>redness</u> at the site of injection and feeling a bit <u>under the weather</u> for a week or two afterwards.

2) You can't have some vaccines if you're <u>already ill</u>, especially if your immune system is weakened.

3) Some people think that immunisation can <u>cause other disorders</u>, e.g. one study <u>suggested</u> a link between the <u>MMR</u> (measles, mumps and rubella) vaccine and <u>autism</u>. Most scientists say the MMR jab is perfectly safe, but a lot of parents aren't willing to take the risk. This has led to a big rise in the number of children catching measles, and some people are now worried about an epidemic.

All in all, the <u>benefits</u> of immunisation normally outweigh the <u>risks</u>. Vaccination is now used all over the world. <u>Smallpox</u> no longer occurs at all, and <u>polio</u> infections have fallen by 99%.

Prevention is better than cure

Science isn't just about doing an experiment, finding the answer and telling everyone about it — scientists often disagree. Not that long ago, scientists had different opinions on the MMR vaccine — and argued about its safety. Many different studies were done before they concluded that it was safe.

Treating Disease — Past and Future

The way we fight disease has changed loads over the last few decades. Thankfully.

Semmelweiss cut deaths by using *antiseptics*

1) While <u>Ignaz Semmelweiss</u> was working in Vienna General Hospital in the 1840s, he saw that women were dying in huge numbers after childbirth from a disease called puerperal fever.

2) He believed that <u>doctors</u> were spreading the disease on their <u>unwashed</u> hands. By telling doctors entering his ward to wash their hands in an <u>antiseptic solution</u>, he cut the death rate from 12% to 2%.

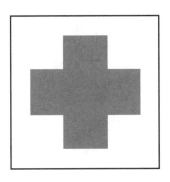

3) The antiseptic solution killed <u>bacteria</u> on doctors' hands, though Semmelweiss didn't know this (the <u>existence</u> of bacteria and their part in causing <u>disease</u> wasn't discovered for another 20 years). So Semmelweiss couldn't <u>prove</u> why his idea worked, and his methods were <u>dropped</u> when he left the hospital (allowing death rates to <u>rise</u> once again — d'oh).

4) Nowadays we know that <u>basic hygiene</u> is essential in controlling disease (though recent reports have found that a lack of it in some <u>modern</u> hospitals has helped the disease <u>MRSA</u> spread — see below).

Antibiotics **changed** the way we **fight infections**

1) <u>Antibiotics</u> were an incredibly important (but accidental) discovery. Some killer diseases (e.g. pneumonia and tuberculosis) suddenly became much easier to treat. The 1940s are sometimes called the era of the <u>antibiotics revolution</u> — it was that big a deal.

2) Unfortunately, bacteria <u>evolve</u> (adapt to their environment). If antibiotics are taken to deal with an infection but not all the bacteria are killed, those that survive may be resistant to the antibiotic and go on to flourish. This process (an example of <u>natural selection</u>) leaves you with an <u>antibiotic-resistant strain</u> of bacteria — not ideal.

3) A good example of antibiotic-resistant bacteria is <u>MRSA</u> (methicillin-resistant *Staphylococcus aureus*) — it's resistant to methicillin, which is one of the most powerful antibiotics around. This is why it's important for patients to always <u>finish</u> a course of antibiotics, and for doctors to avoid <u>over-prescribing</u> them.

Antibiotic resistance is inevitable

Antibiotic resistance is a scary prospect. Bacteria reproduce quickly, and so are pretty fast at evolving to deal with threats (e.g. antibiotics). If we were back in the situation where we had no way to treat bacterial infections, we'd have a nightmare. That's why 'superbugs' like MRSA have been in the news so much. So do your bit, and finish your courses of antibiotics.

Treating Disease — Past and Future

When it comes to treating diseases, viruses are normally a bit more problematic than bacteria...

You *can't* use *antibiotics* to treat *viral* infections

Antibiotics <u>don't</u> destroy viruses.

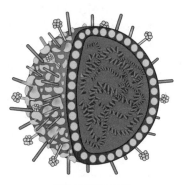

A flu virus

1) Viruses reproduce <u>using your own body cells</u>, which makes it very difficult to develop drugs that destroy just the virus without killing the body cells.

2) <u>Flu</u> and <u>colds</u> are caused by <u>viruses</u>. Usually you just have to wait for your body to deal with the virus, and relieve the <u>symptoms</u> if you start to feel really grotty.

3) There are some <u>antiviral</u> drugs available, but they're usually <u>reserved</u> for very <u>serious</u> viral illnesses (such as AIDS and hepatitis).

We face *new* and *scary dangers* all the time

1) For the last few decades, humans have been able to deal with <u>bacterial infections</u> pretty easily using <u>antibiotics</u>.

2) But there'd be a real problem if a <u>virus</u> or a strain of bacterium evolved so that it was both <u>deadly</u> and could easily pass from <u>person</u> to <u>person</u>. (<u>Flu</u> viruses, for example, evolve quickly so this is quite possible.)

3) If this happened, <u>precautions</u> could be taken to stop the virus spreading in the first place (though this is hard nowadays — millions of people travel by plane every day). And <u>vaccines</u> and <u>antiviral</u> drugs could be developed (though these take <u>time</u> to mass produce).

4) But in the worst-case scenario, a flu <u>pandemic</u> (e.g. one evolved from bird flu) could kill billions of people all over the world.

A pandemic is when a disease spreads all over the world.

Remember — antibiotics DO NOT kill viruses

The recent scare over 'bird flu' shows how vulnerable we are to viral infections. If one came along that was really dangerous and could spread easily, then it'd be a real problem. That's what happened in 1918 when a flu pandemic killed at least 50 million people — that's more than World War 1. Fortunately, nearly all the recent human cases of 'bird flu' have come from direct contact with birds — the virus has rarely passed from person to person.

Warm-Up and Exam Questions

It's easy to think you've learnt everything in the section until you try the warm-up questions.
Don't panic if there's a bit you've forgotten, just go back over that bit until it's firmly fixed in your brain.

Warm-up Questions

1) What is a pathogen?
2) Explain how viruses replicate within your body.
3) Give three functions of white blood cells.
4) All pathogens are parasites. What does this mean?
5) What are antigens?

Exam Questions

1 Immunity can be both active and passive. Describe what is meant by:

(a) active immunity

(1 mark)

(b) passive immunity

(1 mark)

2 The table below describes some of the ways the body defends itself against invading microbes. Match the following defences, **A**, **B**, **C** and **D** with the way they protect the body against invading microbes **1 - 4** in the table.

A Mucus and cilia in the trachea

B Antibodies

C Phagocytes

D Tears

1	Contain the enzyme lysozyme to kill bacteria
2	Engulf and digest foreign molecules
3	Trap bacteria and dust
4	Lock onto invading microbes and mark them out for ingestion by other white blood cells.

(4 marks)

3 (a) Why would a course of antibiotics not be suitable for treatment of flu?

(2 marks)

(b) What might the inappropriate use of antibiotics lead to?

(1 mark)

Warm-Up and Exam Questions

4   Read the following passage.

> Typhoid is an infectious bacterial disease. The typhoid bacterium is often found in food and water where there is poor sanitation. The bacterium causes fever and severe diarrhoea. Typhoid can be fatal but can be treated using antibiotics. Fortunately, the spread of the disease can be reduced by vaccination.

(a)   Explain how vaccinating against typhoid helps reduce the spread of the disease.

(2 marks)

(b)   Put the following stages in order to describe how the cholera vaccine works.

1.   Antibodies attack the typhoid bacteria even though it is harmless.

2.   When live typhoid bacteria infect the body, white blood cells rapidly mass produce antibodies to kill off the pathogen.

3.   Antigens on the dead/inactive bacteria stimulate the production of antibodies by white blood cells.

4.   Dead/inactive typhoid bacteria are injected.

5.   White blood cells remain in the body to provide a memory.

(1 mark)

(c)   Give one short-term side effect that might result from a vaccination.

(1 mark)

5   Some health disorders are listed below. Match the disorders **A**, **B**, **C** and **D** with their causes **1 - 4** in the table.

A   Scurvy

B   Haemophilia

C   Anaemia

D   Diabetes

Disorder	Cause
1	Lack of insulin
2	Lack of iron
3	Lack of vitamin C
4	Inheritance of a faulty gene

(4 marks)

Revision Summary for Section Two

That was a long(ish) section, but kind of interesting, I reckon. These questions will show what you know and what you don't... if you get stuck, have a look back to remind yourself. But before the exam, make sure you can do all of them without any help — if you can't, you know you're definitely <u>not ready</u>.

1) What is "aerobic respiration"? Give the word equation for it.

2) Explain the terms "anaerobic respiration" and "oxygen debt".

3) Name the three main types of digestive enzyme and explain what they do.

4) How does bile help digestion?

5) Name five essential nutrients the body needs and what they're used for.

6)* Put these people in order of how much energy they would likely need from their food (from highest to lowest): a) mechanic, b) professional runner, c) secretary.

7) What's the difference between 'fit' and 'healthy'? Can you be one without being the other?

8) Name five health problems that are associated with obesity.

9) What is cholesterol?

10) State the formula for working out someone's BMI. What is BMI useful for?

11) Why can it be tricky to know how much salt you're eating?

12) Define the terms prescription and addiction.

13) How does a stimulant drug work? Give two examples.

14) Describe the four stages of drug testing.

15) What is a double blind clinical trial?

16) Name a drug that was not tested thoroughly enough and describe the consequences of its use.

17) Describe four different illnesses that smoking can cause.

18) How do carbon monoxide, carcinogens and tar in tobacco smoke each affect the body?

19) Alcohol is a depressant drug. Describe the symptoms of too much alcohol. What effect does alcohol have on the nervous system?

20)*Here is a graph of Mark's blood alcohol concentration against time.

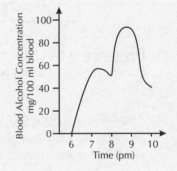

   a)   When did Mark have his first alcoholic drink?

   b)   When did Mark have his second alcoholic drink?

   c)   The legal limit for driving is 80 mg of alcohol per 100 ml of blood. Would Mark have been legally allowed to drive at 9pm?

21) Describe some of the effects that inhaling solvents can cause.

22) What kinds of pain is paracetamol used to relieve? Why shouldn't you exceed the recommended dose?

23) Name two types of painkiller that can only be used under medical supervision.

24) How do aspirin and ibuprofen work? What about opiates?

25) Name the four types of microorganism that cause disease.

26) Explain the difference between benign and malignant tumours.

27) Name the three parts of the body which make up the first line of defence against pathogens.

28) What is the body's second line of defence against pathogens?

29) Explain how immunisation stops you getting infections.

30) Why shouldn't your doctor give you antibiotics for the flu?

*Answers on page 288.

Variation in Plants and Animals

The word 'variation' sounds far too fancy for its own good. All it means is how animals or plants of the same species look or behave slightly differently from each other. You know, a bit taller or a bit fatter or a bit more scary-to-look-at etc. There are two kinds of variation — genetic and environmental.

Genetic variation is caused by genes

1) All animals (including humans) are bound to be slightly different from each other because their genes are slightly different.

2) Genes are the code which determines how your body turns out — they control your inherited traits, e.g. eye colour. We all end up with a slightly different set of genes. The exceptions to this rule are identical twins, because their genes are exactly the same.

Most variation in animals is due to genes and environment

1) Most variation in animals is caused by a mixture of genetic and environmental factors.

2) Almost every single aspect of a human (or other animal) is affected by our environment in some way, however small. In fact it's a lot easier to list the factors which aren't affected in any way by environment:

If you're not sure what "environment" means, think of it as "upbringing" instead.

> 1) Eye colour,
> 2) Hair colour in most animals (in humans, vanity plays a big part),
> 3) Inherited disorders like haemophilia, cystic fibrosis, etc.,
> 4) Blood group.

3) Environment can have a large effect on human growth even before someone's born. For example, a baby's weight at birth can be affected by the mother's diet.

4) And having a poor diet whilst you're growing up can stunt your growth — another environmental variation.

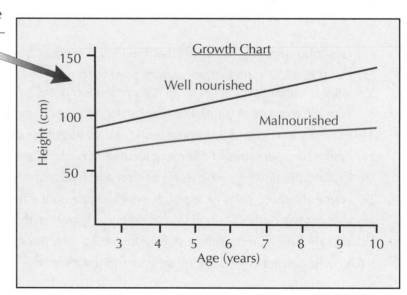

Variation in Plants and Animals

Sometimes it's **hard** to tell which **factor** is **more important**

For some characteristics, it's hard to say which factor is more important — genes or environment...

1) <u>Health</u> — Some people are more likely to get certain <u>diseases</u> (e.g. <u>cancer</u> and <u>heart disease</u>) because of their genes. But <u>lifestyle</u> also affects the risk, e.g. if you smoke or only eat junk food.

2) <u>Intelligence</u> — One theory is that although your <u>maximum possible IQ</u> might be determined by your <u>genes</u>, whether you get to it depends on your <u>environment</u>, e.g. your <u>upbringing</u> and <u>school</u> life.

3) <u>Sporting ability</u> — Again, genes probably determine your <u>potential</u>, but training is important too.

Environmental variation in **plants** is much **greater**

Plants are strongly affected by:

1) sunlight,
2) moisture level,
3) temperature,
4) the mineral content of the soil.

Think about it — if you give your pot plant some plant food (full of lovely minerals), then your plant grows loads faster. Farmers and gardeners use <u>mineral fertilisers</u> to improve crop yields.

For example, plants may grow <u>twice as big</u> or <u>twice as fast</u> due to <u>fairly modest</u> changes in environment such as the amount of <u>sunlight</u> or <u>rainfall</u> they're getting, or how <u>warm</u> it is or what the <u>soil</u> is like.

Most variation is a mixture of genes and environment

So there you go... the "<u>nature versus nurture</u>" debate (*Are you like you are because of the genes you're born with, or because of the way you're brought up?*) summarised in one page. And the winner is... well, <u>both</u> of them. Your genes are pretty vital, but then so is your environment. What an anticlimax.

DNA and Genes

If you're going to get <u>anywhere</u> with this topic you have to make sure you know <u>exactly</u> what <u>DNA</u> is, what and where <u>chromosomes</u> are, and what and where a <u>gene</u> is. If you don't get that sorted out first, then anything else you read about them won't make a lot of sense to you. Whether you're talking about animals or plants, this basic stuff about genes is pretty much the same...

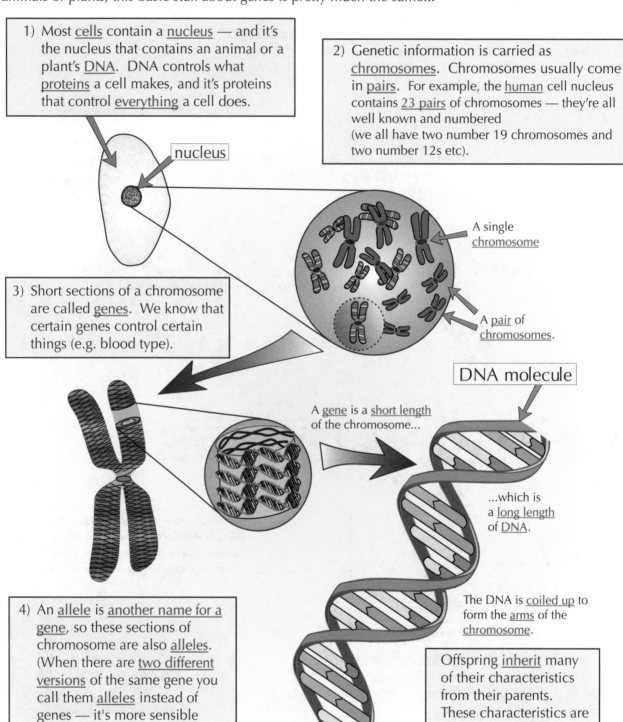

1) Most <u>cells</u> contain a <u>nucleus</u> — and it's the nucleus that contains an animal or a plant's <u>DNA</u>. DNA controls what <u>proteins</u> a cell makes, and it's proteins that control <u>everything</u> a cell does.

nucleus

2) Genetic information is carried as <u>chromosomes</u>. Chromosomes usually come in <u>pairs</u>. For example, the <u>human</u> cell nucleus contains <u>23 pairs</u> of chromosomes — they're all well known and numbered (we all have two number 19 chromosomes and two number 12s etc).

A single <u>chromosome</u>

A <u>pair</u> of <u>chromosomes</u>.

3) Short sections of a chromosome are called <u>genes</u>. We know that certain genes control certain things (e.g. blood type).

DNA molecule

A <u>gene</u> is a <u>short length</u> of the chromosome...

...which is a <u>long length</u> of <u>DNA</u>.

The DNA is <u>coiled up</u> to form the <u>arms</u> of the <u>chromosome</u>.

4) An <u>allele</u> is <u>another name for a gene</u>, so these sections of chromosome are also <u>alleles</u>. (When there are <u>two different versions</u> of the same gene you call them <u>alleles</u> instead of genes — it's more sensible than it sounds!)

Offspring <u>inherit</u> many of their characteristics from their parents. These characteristics are determined by <u>genes</u>.

You need to learn all those technical words

Genes control everything a cell <u>does</u>, as well as what <u>characteristics</u> parents pass on to their kids. It's all to do with <u>proteins</u> — <u>genes</u> control the <u>proteins</u> that are made, and <u>proteins</u> control the cell. The 'rungs' of the DNA 'ladder' are called <u>bases</u>, and there are four different kinds — <u>A</u>, <u>T</u>, <u>C</u> and <u>G</u>. It's the order of these bases in a gene that controls the order that amino acids are strung together in to make a protein.

Asexual Reproduction

Cells can split to form two identical copies of themselves.
This is called <u>mitosis</u> — it's used for growth, repair and asexual reproduction.

Asexual reproduction produces *identical offspring* by *mitosis*

1) A process called <u>mitosis</u> is used in all <u>plants</u> and <u>animals</u> (including you) to <u>grow</u>, and to <u>replace</u> dead or damaged cells.

2) <u>Mitosis</u> involves an existing cell splitting in two to form two offspring <u>identical</u> to the original cell.

3) Some organisms can also <u>reproduce</u> using mitosis, e.g. spider plants, strawberries and potatoes. This is known as <u>asexual</u> reproduction. Here's a <u>definition</u>:

> In <u>asexual reproduction</u> there is only <u>one</u> parent, and the offspring therefore have <u>exactly the same genes</u> as the parent (i.e. they're <u>clones</u> — see p74).

The offspring are **genetically identical** to the parent cell

The really riveting part of mitosis is how the <u>chromosomes split</u> inside the cell. Enjoy...

The DNA is usually all spread out in <u>long strings</u>.

DNA forms into chromosomes. Remember, the <u>double arms</u> are already <u>duplicates</u> of each other.

Chromosomes line up along the centre and then <u>cell fibres pull them apart</u>.

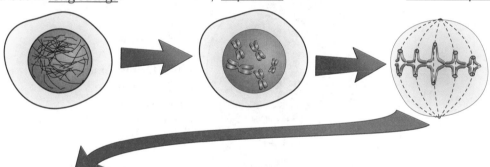

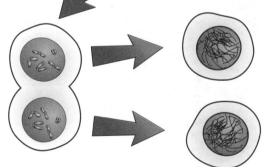

<u>Membranes form</u> around the two sets of chromosome threads. These become the <u>nuclei</u> of the two daughter cells.

The threads <u>unwind</u> into long strands of DNA, which are then <u>duplicated</u>. This means the two new cells have the same number of complete chromosomes as the original cell.

Reproduction doesn't have to involve sex — ask any spider plant

You do need to know all the details on this page, I'm afraid. Certain <u>plants</u> can reproduce asexually, e.g. potatoes, strawberries, daffodils and <u>chlorophytum</u> (spider plants). Spider plants grow tufty bits at the end of their shoots called <u>plantlets</u> — each plantlet is a clone of the original plant. The clone grows roots and becomes a new plant.

Sexual Reproduction and Variation

If you thought reproduction by mitosis was exciting, you'll love this...

Sexual reproduction leads to genetic variation

First, gametes are formed — sperm cells and egg cells...

1) Gametes are <u>sperm cells</u> and <u>egg cells</u>. They're formed in the ovaries or testes from <u>reproductive cells</u>.

2) Human reproductive cells (like all human body cells) have <u>23 pairs</u> of chromosomes. In each pair there's one chromosome that was <u>originally inherited</u> from <u>mum</u>, and one that was inherited from <u>dad</u>.

3) When reproductive cells <u>split</u> into two, some of your dad's chromosomes are grouped with some from your mum. But there'll be no pairs at all now — just one of each of the 23 different types in each of the two new cells. (Each cell therefore has a <u>mixture</u> of your mother's and father's characteristics, but has only half the full number of chromosomes.)

4) These cells form <u>gametes</u> — so gametes have only <u>half</u> the normal amount of genetic information.

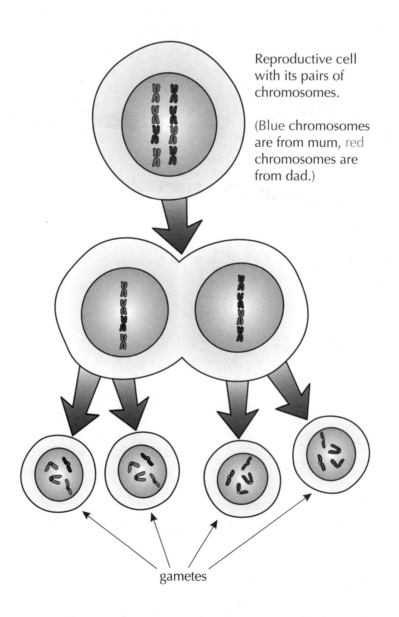

Reproductive cell with its pairs of chromosomes.

(Blue chromosomes are from mum, red chromosomes are from dad.)

gametes

Sexual Reproduction and Variation

So now you know how 46 chromosomes become 23 — here's how they get back to 46...

...then *fertilisation* happens — the *gametes* join together

> Fertilisation is the fusion of male and female gametes, with 23 chromosomes each, forming a zygote (fertilised egg) with 23 pairs of chromosomes.

It's 23 for humans, anyway.

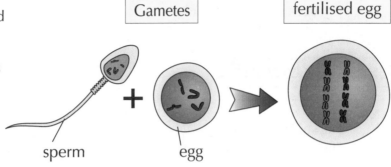

Gametes

fertilised egg

sperm

egg

1) Fertilisation is when the <u>sperm</u> and the <u>egg</u>, with <u>23 chromosomes each</u>, join to form a new cell with the full <u>46 chromosomes</u>. (The 23 chromosomes in one gamete all pair off with their appropriate "partner chromosome" from the other gamete to form the full 23 pairs again.)

2) The fertilised egg then grows by <u>mitosis</u>.

3) The offspring will receive its characteristics as a <u>mixture</u> from the two sets of chromosomes, so it will <u>inherit</u> features from <u>both</u> parents. This is why <u>sexual</u> reproduction produces <u>more variation</u> than <u>asexual</u> reproduction. Pretty cool, eh.

Whenever a cell *copies* itself, *mutations* can happen

A <u>mutation</u> is a genetic change in an organism. It could lead on to some strange new characteristic.

1) Sometimes mutations happen when a cell is <u>splitting</u> (either a <u>reproductive cell</u> splitting to form <u>gametes</u>, or <u>clones</u> forming by <u>mitosis</u>) and something goes <u>wrong</u>.

2) Mutations happen <u>spontaneously</u> — when a chromosome doesn't quite copy itself properly. However, the chance of mutation is <u>increased</u> by exposure to:
   - <u>nuclear radiation</u>, <u>X-rays</u> or <u>ultraviolet</u> light,
   - <u>chemicals</u> called <u>mutagens</u> (<u>cigarette smoke</u> contains <u>mutagens</u>).

3) Mutations are usually harmful.
   - If a mutation occurs in <u>reproductive cells</u>, the offspring might develop <u>abnormally</u> or <u>die</u>.
   - If a mutation occurs in body cells, the mutant cells may start to <u>multiply</u> in an <u>uncontrolled</u> way and <u>invade</u> other parts of the body (which is <u>cancer</u>).

4) <u>Very occasionally</u>, mutations are beneficial and give an organism a <u>survival advantage</u>. The mutation can then spread through the population by <u>natural selection</u> (see pages 92-93).

It should all be starting to come together now

The <u>fusion</u> of male and female <u>gametes</u> to form a <u>zygote</u> — such romance (but this is what makes you different from a spider plant). Remember, variation is important for <u>evolution</u> (p91), and all this applies to <u>any</u> animal or plant that uses sexual reproduction (though the <u>number</u> of chromosomes will be <u>different</u>).

Warm-Up and Exam Questions

Take a deep breath and go through these warm-up questions one by one.
If you don't know these basic facts there's no way you'll cope with the exam questions.

Warm-up Questions

1) Where is DNA found in an animal or plant cell?
2) What is the name given to the type of cell division that produces cells for growth and repair?
3) What are genes?
4) State four environmental factors that can affect the growth of plants.
5) Why does sexual reproduction lead to variation?

Exam Questions

1 The spider plant, *Chlorophytum*, can produce 'baby' spider plants (plantlets) asexually.
The plantlets can be cut off, placed into soil and grown into new spider plants.

Which of the following statements about the new spider plants is true?

A They are genetically different from each other and from the parent plant.

B They are genetically different from each other but the same as the parent plant.

C They are genetically the same as each other but different from the parent plant.

D They are genetically the same as each other and the same as the parent plant.

(1 mark)

2 The diagrams below show four different stages of mitosis.

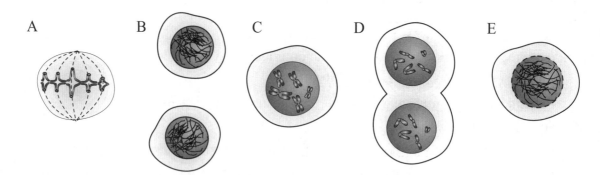

Which of the following correctly shows the order of events during mitosis?

A B E C A D

B C A B E D

C E C A D B

D E D C A B

(1 mark)

Exam Questions

3   Which of the following statements about sperm cells and egg cells is correct?

   A   They contain twice as much genetic information as normal body cells.

   B   They contain the same amount of genetic information as normal body cells.

   C   They do not contain any genetic information.

   D   They contain half the genetic information of normal body cells.

   (1 mark)

4   How many chromosomes does a human liver cell contain?

   A   2

   B   23

   C   23 pairs

   D   46 pairs

   (1 mark)

5   The following piece of advice is taken from a cigarette packet: Smoking Causes Cancer.

   (a)   What is cancer?

   (1 mark)

   (b)   Cancers are the result of mutations. What is a mutation?

   (1 mark)

   (c)   Give two things (other than cigarette smoke) that can increase the risk of mutation.

   (2 marks)

   (d)   Explain why some mutations can be beneficial.

   (2 marks)

6   Andrew and Peter are identical twins. Peter is heavier than Andrew but Andrew is taller.

   (a)   Some people find it hard to tell Andrew and Peter from each other.
         Why do identical twins, such as Andrew and Peter, look very similar?

   (1 mark)

   (b)   Andrew and Peter are not completely identical.
         Explain why Andrew and Peter look slightly different from each other.

   (1 mark)

   (c)   (i)   Give one human characteristic that is determined only by genes.

   (1 mark)

       (ii)   Give one human characteristic that is not affected by genes.

   (1 mark)

Genetic Diagrams

When a <u>single gene</u> controls the inheritance of a characteristic, you can work out the odds of getting it...

Alleles are *different versions* of the *same gene*

1) Most of the time you have <u>two</u> of each gene (i.e. two alleles) — one from each parent.

2) If the alleles are different you have instructions for two different versions of a characteristic (e.g. blue eyes or brown eyes), but you only show one version of the two (e.g. brown eyes). The version of the characteristic that appears is caused by the <u>dominant allele</u>. The other allele is said to be <u>recessive</u>.

3) In genetic diagrams <u>letters</u> are used to represent <u>genes</u>. <u>Dominant</u> alleles are always shown with a <u>capital letter</u> (e.g. 'C') and <u>recessive</u> alleles with a <u>small letter</u> (e.g. 'c').

4) You'll need to know the definitions of homozygous and heterozygous:

Homozygous

If you're <u>homozygous</u> for a trait you have <u>two alleles the same</u> for that particular gene, e.g. CC or cc.

Heterozygous

If you're <u>heterozygous</u> for a trait you have <u>two different alleles</u> for that particular gene, e.g. Cc.

You need to be able to **construct** *and* **explain** *genetic diagrams*

Imagine you're cross-breeding <u>hamsters</u>, and that some have a normal, boring disposition while others have a leaning towards crazy acrobatics. And suppose you know the behaviour is due to one gene...

Let's say that the allele which causes the crazy nature is <u>recessive</u> — so use a '<u>b</u>'.
And normal (boring) behaviour is due to a <u>dominant allele</u> — call it '<u>B</u>'.

1) For an organism to display a <u>recessive</u> characteristic, <u>both</u> its alleles must be <u>recessive</u> — so a crazy hamster must have the alleles 'bb' (i.e. it must be homozygous for this trait).

2) However, a <u>normal hamster</u> could be BB (homozygous) or Bb (heterozygous), because the dominant allele (B) <u>overrules</u> the recessive one (b).

Genetic Diagrams

So if you cross a <u>thoroughbred crazy hamster</u>, genetic type bb, with a <u>thoroughbred normal hamster</u>, BB, you get this:

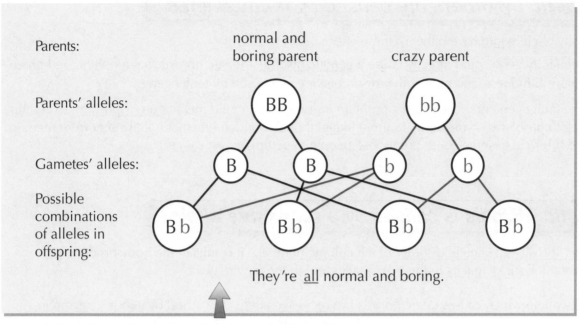

The lines show <u>all</u> the <u>possible</u> ways the parents' alleles <u>could</u> combine.

Remember, only <u>one</u> of these possibilities would <u>actually happen</u> for any one offspring.

When you breed two organisms together to look at one characteristic it's called a MONOHYBRID CROSS.

If two of the offspring from the cross above now breed, they will produce a <u>new combination</u> of offspring:

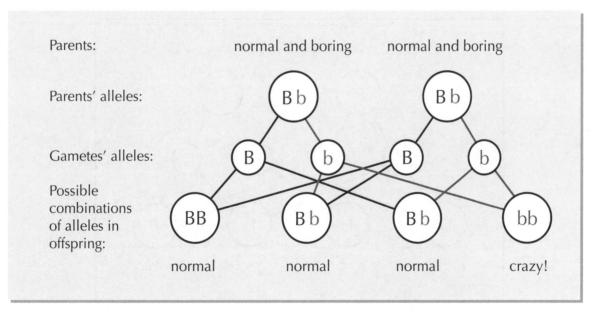

This time, there's a 75% chance of having a normal, boring hamster, and a 25% chance of a crazy one. (To put that another way... you'd expect a 3:1 ratio of normal:crazy hamsters.)

Genetic diagrams can only give you a probability

Interestingly (maybe), you can draw a similar diagram to show the probability of having a <u>boy or a girl</u>. (It's not a single gene that determines sex, but a <u>chromosome</u> — the diagram's exactly the same though.) Women have 2 <u>X-chromosomes</u>, whereas men have <u>an X and a Y</u>, the Y-chromosome being <u>dominant</u>. It turns out the odds are <u>50:50</u>. Have a go at drawing the genetic diagram to show this — or look on p288.

Genetic Disorders

Sometimes an allele might be <u>faulty</u> and <u>not work properly</u>. It can cause more than a few problems...

Genetic disorders are caused by *faulty alleles*

A faulty allele <u>could</u> have either of these effects...

1) The faulty gene could <u>directly</u> cause a <u>genetic disorder</u>. Cystic fibrosis, haemophilia, red-green colour blindness (and many other disorders) are all caused by faulty genes.

2) The faulty gene may not cause a problem in itself, but it could mean a <u>predisposition</u> to certain health problems — meaning it's <u>more likely</u> (though not definite) that you'll suffer from them in the <u>future</u> (e.g. some genes predispose people to getting <u>breast cancer</u>).

Cystic fibrosis is caused by a *recessive allele*

<u>Cystic fibrosis</u> is a <u>genetic disorder</u> of the <u>cell membranes</u>. It results in the body producing a lot of thick sticky <u>mucus</u> in the <u>air passages</u> and in the <u>pancreas</u>.

1) The allele which causes cystic fibrosis is a <u>recessive allele</u>, 'f', carried by about <u>1 person in 25</u>.

2) Because it's recessive, people with only <u>one copy</u> of the allele <u>won't</u> have the disorder — they're known as <u>carriers</u>.

3) For a child to have a chance of inheriting the disorder, <u>both parents</u> must be either <u>carriers</u> or <u>sufferers</u>.

4) As the diagram shows, there's a <u>1 in 4 chance</u> of a child having the disorder if <u>both</u> parents are <u>carriers</u>.

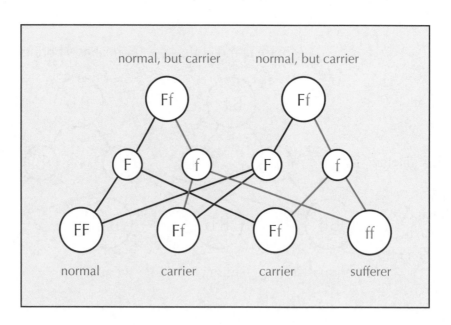

normal, but carrier   normal, but carrier

Ff   Ff

F f F f

FF   Ff   Ff   ff

normal   carrier   carrier   sufferer

That's three genetic diagrams in two pages. Get scribbling

Cystic fibrosis is a terrible condition that affects a lot of people. Luckily for you, you just need to know the symptoms and draw a genetic diagram to show how it's inherited (just like the one above).

Genetic Disorders

There are all sorts of problems with preventing and treating genetic disorders.

Knowing about genetic disorders opens up a **whole can of worms**

Knowing there are inherited conditions in your family raises <u>difficult issues</u>:

- Should all family members be <u>tested</u> to see if they're carriers? Some people might prefer <u>not to know</u>, but is this <u>fair</u> on any partners or future children they might have?

- Is it <u>right</u> for someone who's at risk of passing on a genetic condition to have <u>children</u>? Is it <u>fair</u> to put them under pressure <u>not to</u>, if they decide they want children?

- It's possible to <u>test</u> a foetus for some genetic conditions while it's still in the <u>womb</u>. But if the test is positive, is it right to <u>terminate</u> the pregnancy? The family might not be able to <u>cope</u> with a sick or disabled child, but why should that child have a lesser <u>right to life</u> than a healthy child? Some people think abortion is <u>always wrong</u> under any circumstances.

Gene therapy is being developed to **treat genetic disorders**

Gene therapy means <u>correcting faulty genes</u> — usually a <u>healthy copy</u> of the gene is added.

Example — Treating cystic fibrosis

1) At the moment scientists are trying to cure cystic fibrosis (CF) with gene therapy. One method being trialled is the use of a virus to insert a <u>healthy copy</u> of the gene into cells in the airways.

2) There are still problems — for example, at the moment the effect wears off after a <u>few days</u>. But there are big hopes that gene therapy will one day mean CF can be treated effectively.

3) However, since this kind of gene therapy involves only body cells (and not reproductive cells), the faulty gene would still be passed on to children.

Learn the facts then see what you know

On a related note... the <u>Human Genome Project</u> aimed to map all the genes in a human (see p72).
This has now been completed, and the results could help with future gene therapies. Exciting stuff.

Warm-Up and Exam Questions

There's no better preparation for exam questions than doing... err... practice exam questions.
Hang on, what's this I see...

Warm-up Questions

1) What are alleles?
2) What are genetic disorders caused by?
3) What are the symptoms of the genetic disorder cystic fibrosis?
4) State two genetic disorders (other than cystic fibrosis).
5) If someone has genes that predispose them to breast cancer, will they definitely develop breast cancer?

Exam Questions

1   Handedness is determined by genes.
    The allele for right-handedness, H, is dominant to the allele for left-handedness, h.

    (a)   Complete the genetic diagram below to show the possible combinations
          of alleles in the offspring for the two parents.

                                        parent's alleles

		H	h
parent's alleles	**H**		
	h		

(1 mark)

    (b)   What is the chance of the offspring being left-handed?

          A   25%

          B   50%

          C   75%

          D   100%

(1 mark)

2   Jack found out that his father is a carrier of a genetic disorder. Which of the following is a
    reason why Jack may **not** want to be tested for the same disorder?

    A   Not knowing might be unfair on any partners.

    B   If he is a carrier, he might feel under pressure not to have children.

    C   If he had the genetic disorder he could prepare for its onset.

    D   He could pass on the disorder to his children.

(1 mark)

Exam Questions

3   Which of the following statements about the use of gene therapy to treat cystic fibrosis
     is **not** correct?

   A   At the moment, the effects are only temporary.

   B   It can involve the use of viruses.

   C   It involves inserting a working gene into cells in the airways.

   D   The healthy gene can be passed on to children of the sufferer.

   (1 mark)

4   Cystic fibrosis is a disease caused by recessive alleles.

        F = the normal allele
        f = the faulty allele that leads to cystic fibrosis

   The genetic diagram below shows the possible inheritance of cystic fibrosis from
   one couple.

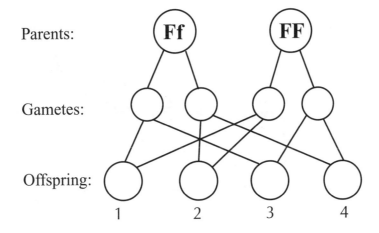

   (a) Complete the genetic diagram.

   (2 marks)

   (b) Which of the possible offspring will be sufferers and which
        will be unaffected?

   (1 mark)

   (c) (i)   What proportion of the possible offspring are homozygous?

   (1 mark)

        (ii) Which of the possible offspring are carriers of the disease?

   (1 mark)

   (d) Explain why someone might want to be tested to see if they are a carrier for
        cystic fibrosis.

   (1 mark)

Genetic Engineering

Scientists can now <u>add</u>, <u>remove</u> or <u>change</u> an organism's <u>genes</u> to alter its characteristics.

Genetic engineering uses enzymes to cut and paste genes

The basic idea is to move <u>useful genes</u> from one organism's chromosomes into the cells of another...

1) A useful gene is "<u>cut</u>" from one organism's chromosome using <u>enzymes</u>.

2) <u>Enzymes</u> are then used to <u>cut</u> another organism's chromosome and to <u>insert</u> the useful gene. This technique is called <u>gene splicing</u>.

3) Scientists use this method to do all sorts of things — for example, the human insulin gene can be inserted into <u>bacteria</u> to <u>produce human insulin</u>:

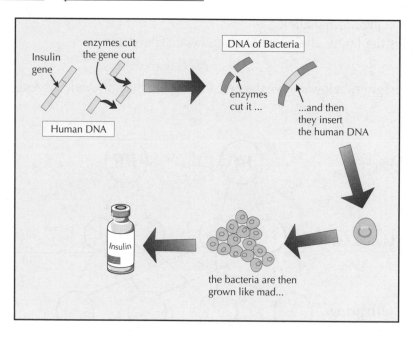

Genes can be transferred into animals and plants

The same method can be used to <u>transfer useful genes</u> into <u>animals</u> and <u>plants</u> at the <u>very early stages</u> of their development (i.e. shortly after <u>fertilisation</u>). This has (or could have) some really <u>useful applications</u>.

1) <u>Genetically modified (GM) plants</u> have been developed that are <u>resistant to viruses</u> and <u>herbicides</u> (chemicals used to kill weeds). And <u>long-life</u> tomatoes can be made by changing the gene that causes the fruit to ripen.

2) Genes can also be inserted into <u>animal embryos</u> so that the animal grows up to have more <u>useful characteristics</u>. For example, <u>sheep</u> have been genetically engineered to produce substances (e.g. drugs) in their <u>milk</u> that can be used to treat <u>human diseases</u>.

'Bt corn' contains a gene from a bacterium that protects against corn borer insects.

3) <u>Genetic disorders</u> like cystic fibrosis are caused by faulty genes. Scientists are trying to cure these disorders by <u>inserting working genes</u> into sufferers. This is called <u>gene therapy</u> — see page 67.

Genetic Engineering

On the face of it, genetic engineering is great. But like most other things, there are benefits and risks that you need to consider.

But genetic engineering is a **controversial** topic...

So, genetic engineering is an exciting new area in science which has the potential for solving many of our problems (e.g. treating diseases, more efficient food production etc.) but not everyone likes it.

1) Some people strongly believe that we shouldn't go tinkering about with genes because it's not natural.

2) There are also worries that changing an organism's genes might accidentally create unplanned problems — which could then get passed on to future generations.

There are **pros** and **cons** with GM crops

Disadvantages

1) Some people say that growing GM crops will affect the number of weeds and flowers (and therefore wildlife) that usually lives in and around the crops — reducing farmland biodiversity.

2) Not everyone is convinced that GM crops are safe. People are worried they may develop allergies to the food — although there's probably no more risk for this than for eating usual foods.

3) A big concern is that transplanted genes may get out into the natural environment. For example, the herbicide resistance gene may be picked up by weeds, creating a new 'superweed' variety.

Advantages

1) On the plus side, GM crops can increase the yield of a crop, making more food.

2) People living in developing nations often lack nutrients in their diets. GM crops could be engineered to contain nutrients that are missing. For example, they're testing 'golden rice' that contains beta-carotene — lack of this substance can cause blindness.

3) GM crops are already being used elsewhere in the world (not the UK), often without any problems.

Genetic engineering has exciting and frightening possibilities

It's up to the Government to weigh up all the evidence before making a decision on how this knowledge is used. All scientists can do is make sure the Government has all the information it needs.

The Human Genome Project

Some people have called the Human Genome Project "more exciting than the first moon landing", or "the search for the Holy Grail of science". Maybe they should get out more.

The idea was to **map** the **25 000** *(or so)* **human genes**

The big idea was to find every single human gene — all 25 000 genes that curl up to form the 23 chromosomes (the other 23 have the same genes — but maybe different versions).
Well... they've found them all — and now they're trying to figure out what each one does.

> In the exam, they'll probably ask you to say what's good about it, what's bad about it, or both.

The **good stuff** — *improving* **medicine** *and* **forensic science**

1) Predict and prevent diseases

If doctors knew what genes predisposed people to what diseases, we could all get individually tailored advice on the best diet and lifestyle to avoid our likely problems. Better still, cures could be found for genetic diseases like cystic fibrosis (p66) and sickle cell anaemia.

2) Develop new and better medicines

Maybe one day we'll all have medicines designed especially for us.

3) Accurate diagnoses

Some diseases are hard to test for (e.g. you can only tell for sure if someone has Alzheimer's after they die), but if we know the genetic cause, accurate testing will be a lot easier.

4) Improve forensic science

Forensic scientists can produce a 'DNA fingerprint' from material found at a crime scene. If this matches a suspect's DNA, he or she was almost certainly there. One day it might be possible to figure out what a suspect looks like from DNA found at the scene of a crime (e.g. eye, hair and skin colour).

The Human Genome Project

Just like genetic engineering, the sequencing of the human genome brought up loads of ethical problems and uncertainties. And luckily for you, you need to know about them all...

The *bad stuff* — it could be a *scary world* if you're *not perfect*

Knowing so much about our own genes could end up giving us all more things to worry about.

1) Increased stress

If someone knew from an early age that they're susceptible to a nasty brain disease, they <u>could</u> panic every time they get a headache (even if they never get the disease).

2) Gene-ism

People with genetic problems <u>could</u> come under pressure not to have children.

3) Discrimination by employers and insurers

Life insurance <u>could</u> become impossible to get (or blummin' expensive) if you have any genetic likelihood of serious disease. And employers may want to discriminate against people too.

'Designer babies' mean parents choose characteristics

1) In <u>theory</u> it might one day be possible to genetically engineer a <u>human embryo</u> to have all the genes that you want. In <u>reality</u>, it's <u>very</u> tricky, and won't happen for a while (if ever).

2) It's far more realistic to use 'screening' instead. This involves using <u>IVF</u> and <u>checking</u> an embryo for certain genetic defects <u>before</u> implanting it in the womb (see page 16 for more on IVF).

3) Scientists can also screen embryos by <u>sex</u>. At the moment, it's <u>illegal</u> (in most cases) to screen embryos by sex just because parents <u>want</u> a boy or a girl. But if there's a family history of a genetic disorder that only <u>males</u> can suffer from (e.g. haemophilia), a couple can choose to have a <u>girl</u>.

4) Parents may also be allowed to screen embryos for a '<u>tissue match</u>' if they have an ill child and a new child could provide tissue for transplant (but some people worry the new child may feel like a 'spare part factory').

Knowing too much about our genes could lead to discrimination

These are only <u>possibilities</u> — some may happen <u>soon</u>, some will take <u>ages</u>, and others might not happen at all. Remember... anything to do with genetics tends to be <u>controversial</u>. But before you can come to a sensible decision about the <u>ethics</u> of it all, you need to know the <u>facts</u>.

Cloning

Nature makes clones using mitosis (see page 59). This page is about how humans make them.

Plants *can be cloned from* cuttings *and by* tissue culture

Cuttings

1) Gardeners can take <u>cuttings</u> from good parent plants, and then plant them to produce <u>genetically identical copies</u> (clones) of the parent plant.

2) These plants can be produced <u>quickly and cheaply</u>.

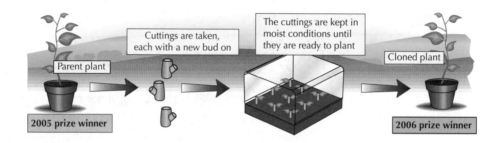

Parent plant — 2005 prize winner

Cuttings are taken, each with a new bud on

The cuttings are kept in moist conditions until they are ready to plant

Cloned plant — 2006 prize winner

Tissue culture

This is where <u>a few plant cells</u> are put in a <u>growth medium</u> with <u>hormones</u>, and they then grow into <u>new plants</u> — <u>clones</u> of the parent plant. The <u>advantages</u> of using <u>tissue culture</u> are that you can make new plants very <u>quickly</u>, in very little <u>space</u>, and you can <u>grow them all year</u>.

The <u>disadvantage</u> to both these methods is a '<u>reduced gene pool</u>' (see below).

You can make animal *clones using* embryo transplants

Farmers can produce <u>cloned offspring</u> from their best bull and cow — using <u>embryo transplants</u>.

1) <u>Sperm cells</u> are taken from a prize bull, and <u>egg cells</u> from a prize cow. The sperm are then used to <u>artificially fertilise</u> an egg cell. The <u>embryo</u> that develops is then <u>split</u> many times (to form <u>clones</u>).

2) These <u>cloned embryos</u> can then be <u>implanted</u> into lots of other cows where they grow into <u>baby calves</u> (which will all be <u>genetically identical</u> to each other).

3) The advantage of this is that <u>hundreds</u> of "ideal" offspring can be produced <u>every year</u> from the best bull and cow.

4) The big disadvantage (as usual) is a <u>reduced gene pool</u>.

Cloning reduces *the* gene pool

1) A "<u>reduced gene pool</u>" means fewer alleles in a population — which will happen if you breed from the <u>same</u> plants or animals all the time.

2) If a population are all closely <u>related</u> and a new disease appears, all the plants or animals could be wiped out — there may be <u>no</u> <u>allele</u> in the population giving <u>resistance</u> to the disease.

Oh Eck.

Cloning

There's a second way to clone animals, and it's even more controversial than the first...

Adult cell cloning is another way to make a clone...

1) Adult cell cloning is the technique that was used to create Dolly — the world-famous cloned sheep.

2) Dolly was made by taking a sheep egg cell and removing its genetic material. A complete set of chromosomes from the cell of an adult sheep was then inserted into the 'empty' egg cell, which then grew into an embryo. This eventually grew into a sheep that was genetically identical to the original adult.

3) Human adult cell cloning could be used to help treat various diseases. A cloned embryo that is genetically identical to the sufferer could be created and embryonic stem cells extracted from it. (These can become any cell in the body and could be used to grow replacement cells or organs — without fear of them being rejected by the sufferer's immune system.)

4) Some people think it's unethical to do this as embryos genetically identical to the sufferer are created and then destroyed.

Adult cell cloning

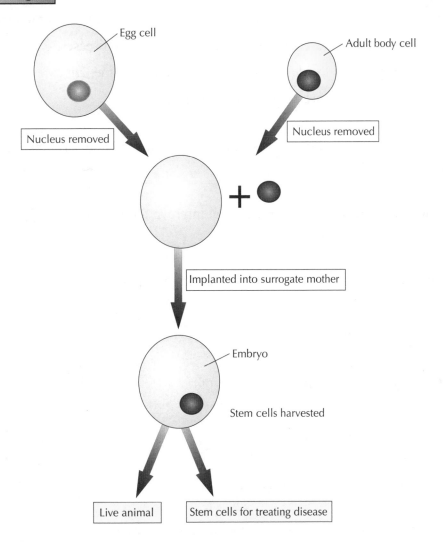

Cloning produces genetically identical organisms

Cloning can be a controversial topic — especially when it's to do with cloning animals (and especially humans). Is it healthy scientific progress, or are we trying to 'play God'?

Warm-Up and Exam Questions

By doing these warm-up questions, you'll soon find out if you've got the basic facts straight. If not, you'll really struggle, so take the time to go back over the bits you don't know.

Warm-up Questions

1) What is the human genome?
2) What is meant by the term 'designer baby'?
3) What is gene splicing?
4) What is a clone?
5) Name two ways that plant clones can be produced.

Exam Questions

1   Many people disagree over the social impacts of the human genome project.

(a) Which one of the following is **not** an advantage of mapping the human genome?

    A   People with genetic problems could come under pressure not to have children.

    B   It may allow more doctors to predict and prevent diseases.

    C   It may enable the development of more effective medicines.

    D   It may allow more accurate diagnoses.

(1 mark)

(b) Describe how mapping the human genome could lead to discrimination by insurance companies.

(1 mark)

(c) How might the human genome project improve forensic science?

(2 marks)

2   The advantage of using tissue culture to produce new plants is that

    A   it is expensive

    B   useful genes can be added

    C   defective genes can be removed

    D   new plants can be produced very quickly

(1 mark)

3   Cloning organisms results in a reduced gene pool.

(a) What is meant by 'reduced gene pool'?

(1 mark)

(b) Describe the problems associated with a reduced gene pool.

(1 mark)

Exam Questions

4 Farmers can produce cloned offspring from their best bull and cow using embryo transplants.

    (a) Why do farmers use this procedure?

(1 mark)

    (b) Describe the process of embryo transplantation.

(4 marks)

5 Organisms can be genetically modified.

    (a) Explain what is meant by 'genetically modified'.

(2 marks)

    (b) Give three functions of enzymes used in genetic engineering.

(3 marks)

    (c) Suggest one useful way that plants can be genetically modified.

(1 mark)

    (d) Suggest one useful way that animals can be genetically modified.

(1 mark)

    (e) Some people think that it is wrong to genetically modify plants.
         Give two different objections that people might have.

(2 marks)

6 In 1997 scientists the Roslin Institute issued a press release to tell the world about the birth of Dolly, a sheep that had been cloned using adult cells.

    (a) Explain how Dolly was made using adult cell cloning.

(3 marks)

    (b) Human adult cell cloning could be used to help treat various diseases.

        (i) Suggest how this could be done.

(2 marks)

        (ii) Explain why some people are opposed to human adult cell cloning.

(1 mark)

Photosynthesis

Bit of a deviation from the 'genetics' theme here... but try not to worry about that too much.

Photosynthesis produces glucose from sunlight

1) Photosynthesis uses energy from the Sun to change carbon dioxide and water into glucose and oxygen.

carbon dioxide + water $\xrightarrow[\text{chlorophyll}]{\text{LIGHT}}$ glucose + oxygen

$$6CO_2 + 6H_2O \xrightarrow[\text{chlorophyll}]{\text{LIGHT}} C_6H_{12}O_6 + 6O_2$$

Photosynthesis and respiration are opposite processes — see p24 for respiration.

2) It takes place in the chloroplasts of plant cells.
Chloroplasts contain chlorophyll that absorbs the light energy.

Glucose is converted into other substances

Glucose is soluble, which makes it good for transporting to other places in the plant. It's also a small molecule — so it can diffuse in and out of cells easily. Here's how plants use the glucose they make:

1) For respiration

Plants use some of the glucose for respiration. This releases energy so they can convert the rest of the glucose into various other useful substances.

2) Making cell walls

Glucose is converted into cellulose for making cell walls, especially in a rapidly growing plant.

3) Stored in seeds

Glucose is turned into lipids (fats and oils) for storing in seeds. Sunflower seeds, for example, contain a lot of oil — we get cooking oil and margarine from them.

4) Stored as starch

Glucose is turned into starch and stored in roots, stems and leaves, ready for use when photosynthesis isn't happening, like at night. (Starch is insoluble, which makes it much better for storing — it doesn't bloat the storage cells by drawing water in like glucose would.)

5) Making proteins

Glucose is combined with nitrates (collected from the soil) to make amino acids, which are then made into proteins. These are used for growth and repair.

Photosynthesis

There's always something slowing plants down...

Three limiting factors control the rate of photosynthesis

1) Not enough light slows down the rate of photosynthesis

1) If the <u>light level</u> is raised, the rate of photosynthesis will <u>increase</u>, but only up to a <u>certain point</u>.

2) Beyond that, it won't make any <u>difference</u> because then it'll be either the <u>temperature</u> or the CO_2 level which is now the limiting factor.

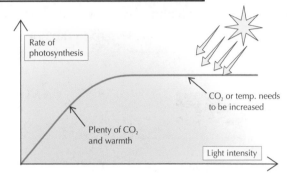

2) Too little carbon dioxide also slows it down

1) As with light intensity, the amount of CO_2 will only <u>increase</u> the rate of photosynthesis up to a point.

2) After this the graph <u>flattens out</u>, showing that CO_2 is no longer the limiting factor.

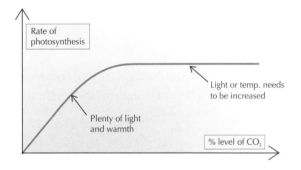

3) The temperature has to be just right

1) As the <u>temperature increases</u>, so does the <u>rate</u> of photosynthesis.

2) But if the <u>temperature</u> is too high, the plant's <u>enzymes</u> will be <u>destroyed</u>, so the rate rapidly <u>decreases</u>. This happens at about <u>45 °C</u>.

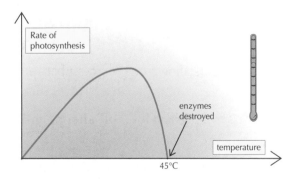

Convert this page into stored information

<u>Photosynthesis</u> is important. All the energy we get from eating comes from it. When we eat <u>plants</u>, we're consuming the energy they've made, and when we eat <u>meat</u>, we're eating animals who got their energy from eating plants, or from eating animals that have eaten other animals who have... and so on.

Warm-Up and Exam Questions

Without a good warm-up you're likely to strain a brain cell or two. So take the time to run through these simple questions and get the basic facts straight before plunging into exam questions.

Warm-up Questions

1) Where in a plant leaf cells does photosynthesis take place?
2) What is the name of the green pigment that is important in photosynthesis?
3) Give the word and balanced symbol equation for photosynthesis.
4) How does the concentration of carbon dioxide affect the rate of photosynthesis?

Exam Questions

1   Glucose is an ideal transport molecule. This is because it is

    A   small and insoluble

    B   large and soluble

    C   small and soluble

    D   large and insoluble

(1 mark)

2   Toby and Jan carry out an experiment to investigate how light intensity affects the rate of photosynthesis. The graph below shows their results.

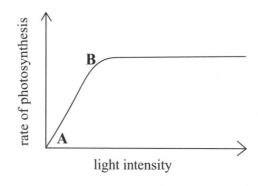

(a)   What trend is shown on the graph between points A and B?

(1 mark)

(b)   Why does the graph level off after point B?

(2 marks)

(c)   Give one thing that plants use the products of photosynthesis for.

(1 mark)

(d)   Some of the glucose made in photosynthesis is converted to starch for storage.
      Explain why starch is such a good storage molecule.

(2 marks)

SECTION THREE — GENETICS AND EVOLUTION

Ecosystems

An <u>ecosystem</u> is all the <u>different organisms</u> living together in a <u>particular environment</u>. Sounds cosy.

Artificial ecosystems can be carefully controlled

1) There are <u>two types of ecosystem</u> you need to know about:

> A <u>natural ecosystem</u> is one where humans <u>don't control the processes</u> going on within it.
>
> An <u>artificial ecosystem</u> is one where humans <u>deliberately</u> promote the growth of certain living organisms and get rid of others which threaten their well-being.

2) Humans <u>might</u> affect <u>natural ecosystems</u> in some way, but they <u>don't take deliberate steps</u> to decide what animals and plants should be there.

3) <u>Artificial ecosystems</u> are most common in money-making enterprises, e.g. <u>farms</u> and <u>market gardens</u>. Things like weedkillers, pesticides and fertilisers are used to control conditions. <u>Artificial ecosystems</u> normally have a <u>smaller number of species</u> (less <u>biodiversity</u>) than natural ones.

Estimate population sizes in an ecosystem using a quadrat

A <u>quadrat</u> is a square frame enclosing a known area.
You just place it on the ground, and look at what's inside it.

To estimate <u>population size</u>:

1) Count all the organisms in a <u>1 m² quadrat</u>.

2) Multiply the number of organisms by the <u>total area</u> (in m²) of the habitat.

3) Er, that's it. Bob's your uncle.

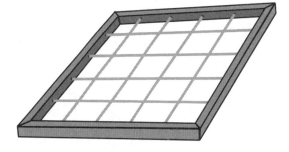

A quadrat

Two important points about this kind of counting method...

1) The <u>sample size</u> affects the <u>accuracy</u> of the estimate — the bigger your sample, the more accurate your estimate of the total population is likely to be. (So it'd be better to use the quadrat more than once, get an <u>average</u> value for the number of organisms in a 1 m² quadrat, then multiply that by the total area.)

2) The sample may not be <u>representative</u> of the population, i.e. what you find in your sample might be different from what you'd have found if you'd looked somewhere else.

Ecosystems and Species

Organisms are <u>classified</u> depending on how <u>closely related</u> they are to other groups of organisms. It's just sometimes quite <u>difficult</u> to work out how closely related two groups are...

Different **organisms** belong to different **species**

1) Organisms are of the <u>same species</u> if they can <u>breed</u> to produce <u>fertile offspring</u>.

2) If you interbreed a male from one species with a female from a <u>different</u> species you'll get a <u>hybrid</u> (that's if you get anything at all). For example, a <u>mule</u> is a cross between a donkey and a horse. But hybrids are <u>infertile</u> so they <u>aren't</u> new species.

Unrelated species may have **similar features**

1) Similar species often share a <u>recent common ancestor</u>, so they're <u>closely related</u> in evolutionary terms. They often look <u>alike</u> and tend to live in similar types of <u>habitat</u>, e.g. whales and dolphins.

2) This isn't always the case though — closely related species may look <u>very different</u> if they have evolved to live in <u>different habitats</u>, e.g. llamas and camels.

3) Species that are <u>very different genetically</u> may also end up looking alike. E.g. dolphins and sharks look pretty similar and swim in a similar way. But they're totally different species — dolphins are <u>mammals</u> and sharks are <u>fish</u>.

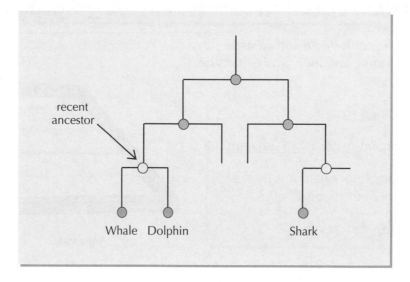

4) So to explain the similarities and differences between species, you have to consider how they're related in <u>evolutionary terms</u> AND the <u>type of environment</u> they've adapted to survive in.

Ecosystems — aren't they used in submarines

It's possible to breed lions and tigers together... it's true — they produce <u>hybrids</u> called tigons and ligers. They look a bit like lions and a bit like tigers... as you'd expect. In the same way, a bat is (I think) just a hybrid of a bird and a cat. And a donkey is the result of breeding a dog and a monkey.

Pyramids of Biomass and Food Chains

<u>Ecosystems</u> — communities of things living side by side in harmony... before eating each other. And so on to <u>food chains</u>.

Pyramids of biomass show weight

1) <u>Biomass</u> is how much the creatures at each level of a food chain would <u>weigh</u> if you <u>put them together</u>.

2) The pyramid of biomass below shows the <u>food chain</u> of a mini meadow ecosystem. The dandelions are the <u>provider</u> (starting point) — they're eaten by the rabbits (primary consumers), which are eaten by the fox (secondary consumer)... and so on.

3) If you weighed them, all the <u>dandelions</u> would have a <u>big biomass</u> and the <u>hundreds of fleas</u> would have <u>a very small biomass</u>. Biomass pyramids are <u>always a pyramid shape</u>.

4) Each time you go <u>up</u> one level (one <u>trophic level</u> if you fancy showing off), the mass of organisms goes <u>down</u>. It takes a lot of food from the level below to keep any one animal alive.

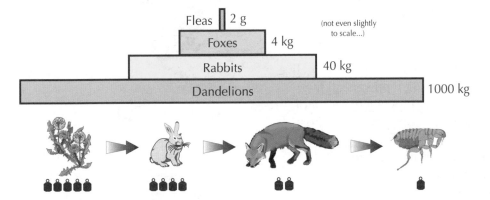

Fleas 2 g (not even slightly to scale...)
Foxes 4 kg
Rabbits 40 kg
Dandelions 1000 kg

It'd be a pain to pull up <u>every single dandelion</u> in an ecosystem and weigh the lot. So, you'd pull up all the dandelions in (say) 1 m² and weigh them all, to give the biomass of dandelions <u>per m²</u>. Some biomass measurements are <u>dry mass</u>, i.e. you'd <u>dry out</u> organisms before weighing them. To save biologists having to <u>kill a fox and dry it out</u> every time they want to measure fox biomass, the <u>average dry mass per fox</u> is available in handy reference resources.

Don't forget that if you get a question on <u>biomass</u> in the exam you need to include the <u>units</u>.

Pyramids of number are less useful

1) <u>Pyramids of number</u> show the <u>number</u> of organisms at each level (usually, numbers <u>fall</u> as you go <u>up</u> the food chain).

2) Sometimes they're <u>not</u> pyramid-shaped though — that's why biomass pyramids are better.

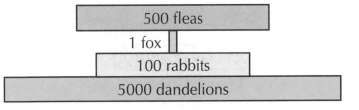

500 fleas
1 fox
100 rabbits
5000 dandelions

Pyramids of Biomass — the eighth wonder of the world

Pyramids of biomass are a way of describing food chains <u>quantitatively</u> (rather than just saying 'foxes eat rabbits', you say <u>how many foxes</u> eat <u>how many rabbits</u>, etc.). The 'minimise lost energy' thing is what gives you <u>battery farming</u>... if you stop the poor hens running around, less energy is lost. Hmm...

Pyramids of Biomass and Food Chains

So if there's so much <u>mass</u> to begin with, how come you end up with two grammes of fleas?

All that **energy** just **disappears** somehow...

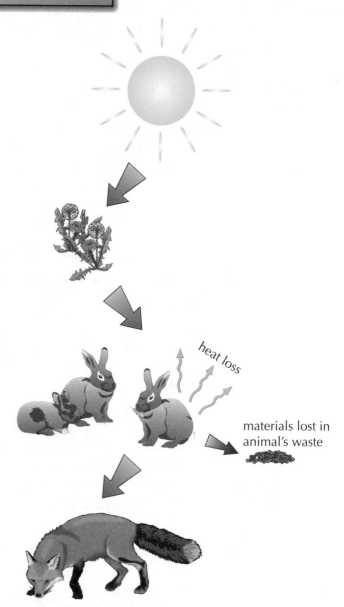

1) Energy from the <u>Sun</u> is the <u>source of energy</u> for nearly <u>all life on Earth</u>.

2) <u>Plants</u> convert <u>a small %</u> of the light energy that falls on them <u>into glucose</u>. The <u>rabbit</u> then <u>eats</u> the <u>plant</u>. It <u>uses up</u> some of the energy it gets from the plant — some of the rest is <u>stored</u> in its body. Then the <u>fox eats</u> the <u>rabbit</u> and gets some of the energy stored in the rabbit's body.

3) Energy is used up at each stage to <u>stay alive</u>, i.e. in <u>respiration</u>, which powers <u>all life processes</u>, including <u>movement</u>. A lot of energy is <u>lost to the surroundings</u> as <u>heat</u>. This is especially true for <u>mammals and birds</u>, whose bodies must be kept at a <u>constant temperature</u> — normally higher than their surroundings.

heat loss

materials lost in animal's waste

4) <u>Material and energy</u> are also lost from the food chain in <u>droppings</u> — if you set dried droppings alight they burn, proving they still have chemical energy in them.

5) This energy is said to be '<u>lost</u>' — it doesn't actually disappear but the next animal in the food chain can't use it.

Food-production efficiency

- For a <u>given area</u> of land, you can produce <u>more food</u> (for humans) by growing <u>crops</u> rather than by <u>grazing animals</u>. This is because you are reducing the number of stages in the food chain.

- But some land is <u>unsuitable</u> for growing crops, like <u>moorland</u> or <u>fellsides</u>. In these places, animals like <u>sheep</u> and <u>deer</u> are often the <u>best</u> way to get food from the land.

Competition and Populations

Organisms have to <u>compete for resources</u> in the environment where they live.

Population size is limited by available resources

<u>Population size</u> is limited by:

1) The <u>total amount of food</u> or nutrients available
   (plants don't eat, but they get <u>minerals</u> from the soil).

2) The amount of <u>water</u> available.

3) The <u>amount of light available</u> (this applies only to plants really).

4) The quality and amount of <u>shelter</u> available.

Animals and plants of the <u>same</u> species and of <u>different</u> species will <u>COMPETE</u> against each other for these resources. They all want to <u>survive</u> and <u>reproduce</u>.

<u>Similar organisms</u> will be in the <u>closest competition</u> — they'll be competing for the same <u>ecological niche</u>.

Example: red and grey squirrels

1) These two different species like the same kind of <u>habitat</u>, same kind of <u>food</u>, type of <u>shelter</u>, etc.

2) Grey squirrels are <u>better adapted</u> to <u>deciduous woodland</u>, so when they were introduced into Britain, red squirrels <u>disappeared</u> from many areas — they just couldn't compete.

Populations of prey and predators go in cycles

In a community containing <u>prey</u> and <u>predators</u> (as most of them do of course):

1) The <u>population</u> of any species is usually <u>limited</u> by the amount of <u>food</u> available.

2) If the population of the <u>prey</u> increases, then so will the population of the <u>predators</u>.

3) However as the population of predators <u>increases</u>, the number of prey will <u>decrease</u>.

E.g. <u>More grass</u> means <u>more rabbits</u>. More rabbits means <u>more foxes</u>. But more foxes means <u>less rabbits</u>. Eventually less rabbits will mean <u>less foxes again</u>. This <u>up and down pattern</u> continues...

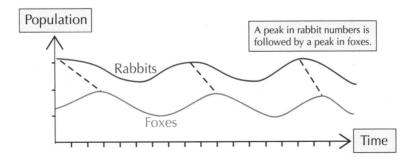

A peak in rabbit numbers is followed by a peak in foxes.

Competition and Populations

All organisms <u>depend</u>, at least to some extent, on other organisms in their <u>ecosystem</u>.
There are some species that take this to the <u>extreme</u>.

Parasites *and* mutualistic *relationships*

The <u>survival</u> of some organisms can <u>depend</u> almost entirely on the presence of <u>other species</u>.

Parasites live off a **host**

Parasites <u>take</u> what they need to survive, <u>without</u> giving anything <u>back</u>.
This often <u>harms</u> the host — which makes it a win-lose situation.

Examples:

- <u>Tapeworms</u> absorb lots of <u>nutrients</u> from the host, causing them to suffer from <u>malnutrition</u>.

- <u>Fleas</u> are parasites. Dogs gain nothing from having fleas (unless you count hundreds of bites).

Mutualism is **beneficial**

Mutualism is a relationship where <u>both</u> organisms benefit — so it's a win-win relationship.

Examples:

- Most plants have to rely on <u>nitrogen-fixing bacteria</u> in the soil to get the <u>nitrates</u> that they need. But <u>leguminous plants</u> carry the bacteria in <u>nodules</u> in their <u>roots</u>. The bacteria get a constant supply of <u>sugar</u> from the plant, and the plant gets essential <u>nitrates</u> from the bacteria.

- '<u>Cleaner species</u>' are fantastic. E.g. <u>oxpeckers</u> live on the backs of <u>buffalo</u>. Not only do they <u>eat pests</u> on the buffalo, like ticks, flies and maggots (providing the oxpeckers with a source of food), but they also <u>alert</u> the animal to any <u>predators</u> that are near, by hissing.

Revision stress — don't let it eat you up

In the exam you might get asked about the distribution of <u>any</u> animals or plants. Just think about
what they would need to survive. And remember, if things are in <u>limited supply</u> then there's going to
be <u>competition</u>. And the more similar the needs of the organisms, the more they'll have to compete.

Warm-Up and Exam Questions

It's easy to think you've learnt everything in the section until you try the warm-up questions. Don't panic if there are bits you've forgotten. Just go back over them until they're firmly fixed in your brain.

Warm-up Questions

1) What is a quadrat and what is it used for?
2) What is the difference between a natural ecosystem and an artificial ecosystem?
3) State three factors that limit the population size for any organism.
4) What is the source of energy for nearly all life on Earth?
5) Why is it usually more efficient to grow crops than to raise livestock for food?

Exam Questions

1   Which of the following is an example of mutualism?

   A   Tapeworms absorbing nutrients from its host.

   B   Bacteria fixing nitrogen for plants and obtaining a supply of sugar from the plant.

   C   Fleas feeding on the blood of a vertebrate host.

   D   Foxes depending on rabbits for a source of nutrients.

(1 mark)

2   Which of the following statements is **not** true about organisms of the same species?

   A   They are able to breed to produce a sterile hybrid.

   B   They live in similar types of habitat.

   C   They are able to breed to produce fertile offspring.

   D   They look similar.

(1 mark)

3   (a)   A dandelion is a producer in a food chain. What are producers?

   A   Organisms that obtain their energy from soil and make their own food.

   B   Organisms that obtain their energy from a primary consumer.

   C   Organisms that obtain their energy from light and make their own food.

   D   The second level of the food chain.

(1 mark)

   (b)   Students counted 22 dandelions in a 5 m^2 area of a 30 m^2 field.
   Estimate the population size of dandelions in the whole field.

(2 marks)

Exam Questions

4   The graph below shows the changes in a population of owls and in a population of mice over a number of years in an area.

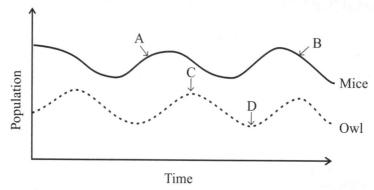

(a) At which point are there fewest owls?

(1 mark)

(b) Suggest why the number of owls falls after point C?

(2 marks)

(c) Suggest what might happen to the owl population if another species was introduced that also eats mice as its main food source?

(1 mark)

5   The diagram below shows a pyramid of biomass for the organisms in a woodland food chain.

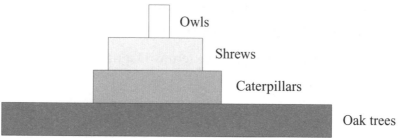

(a) What is biomass?

(1 mark)

(b) Why does biomass decline at each stage in the food chain?

(2 marks)

(c) Why do mammals and birds loose a lot of energy as heat?

(1 mark)

(d) Give one other way in which energy is lost from the food chain.

(1 mark)

(e) What name is given to the different levels in the pyramid?

(1 mark)

Fossils

From a scientific point of view, we're really <u>lucky</u> that fossils exist. They can provide so much <u>information</u> about <u>evolution</u> and the <u>history of life</u> on our planet.

Fossils provide lots of evidence for evolution

A fossil is <u>any trace</u> of an animal or plant that lived long ago. They show how today's species have <u>changed</u> and <u>developed</u> over millions of years. There are <u>three</u> ways fossils can be formed:

1) From gradual replacement by minerals

(Most fossils happen this way.)

1) Things like <u>teeth</u>, <u>shells</u>, <u>bones</u>... which <u>don't decay</u> easily, can last ages when <u>buried</u>.

2) They're eventually <u>replaced by minerals</u>, forming a <u>rock-like substance</u> shaped like the original hard part.

3) The fossil stays <u>distinct</u> inside rock, and is eventually <u>dug up</u>.

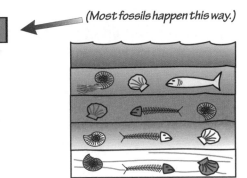

2) From casts and impressions

1) Sometimes, fossils form when an organism's buried in a <u>soft</u> material like <u>clay</u>.

2) The clay hardens around it and the organism <u>decays</u>, leaving a <u>cast</u> of itself.

3) Things like <u>footprints</u> can be pressed into these materials when soft, leaving an <u>impression</u> when it hardens.

3) From preservation in places where no decay happens

The <u>whole</u> original plant or animal may survive for <u>thousands of years</u>:

a) In <u>amber</u> — no <u>oxygen</u> or <u>moisture</u> for the <u>decay microbes</u>.

<u>Insects</u> are often found fully preserved in amber, which is a clear yellow "stone" made of <u>fossilised resin</u> that ran out of an ancient tree millions of years ago, engulfing the insect.

b) In <u>glaciers</u> — too <u>cold</u> for the <u>decay microbes</u> to work.

Apparently, a <u>woolly mammoth</u> was found fully preserved in a glacier somewhere several years ago.

c) In <u>waterlogged bogs</u> — too <u>acidic</u> for <u>decay microbes</u>.

A <u>10 000 year old man</u> was found in a bog a few years ago. He was dead, and a bit squashed but otherwise quite well preserved, although it was clear he'd been <u>murdered</u>.

Fossils

Fossils are definitely a great source of information, but there are some gaps that you have to fill in based on the best available evidence.

The **fossil record** is **incomplete**

1) Fossils found in rock layers tell us two things

i) What the creatures and plants looked like.

ii) How long ago they existed. Generally, the deeper the rock, the older the fossil (though rocks do get pushed upwards, so old rocks can become exposed).

2) There are 'missing links' in the fossil record

This is because very few dead plants or animals actually turn into fossils. Most just decay away. There are fossils yet to be discovered that might help complete the picture.

3) The fossil record of the horse provides strong evidence for evolution

If you look at fossilised bones, you can put together a family tree of the horse — showing how modern horses have evolved.

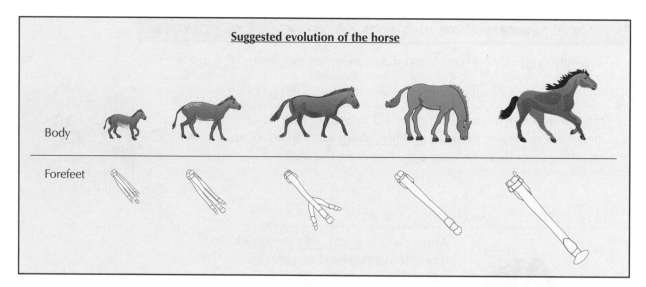

Suggested evolution of the horse

Body

Forefeet

Over many generations, the middle toe has got slowly bigger and bigger to form the familiar hoof of today's horse, and the animals have got generally larger. But making an accurate family tree is tricky. Not all the similar-looking fossils found are ancestors of today's horses — some were 'evolutionary dead-ends'.

Don't get bogged down by all this information

The fossil record provides good evidence for evolution, but it can't prove it. But proving a theory of something that happens over millions of years was never going to be straightforward, I guess.

Evolution

The <u>theory of evolution</u> states that one of your (probably very distant) ancestors was a <u>blob</u> in a swamp somewhere. Something like that, anyway.

*Evolution is a **slow** process*

1) This suggests that all the animals and plants on Earth gradually 'evolved' over <u>millions</u> of years, rather than just suddenly popping into existence. Makes sense.

2) Life on Earth began as <u>simple organisms</u> from which all the more complex organisms evolved. And it only took about <u>3 000 000 000 years</u>.

*There are lots of **modern examples** of evolution*

*1) **Peppered moths** adapted their colour*

<u>Peppered moths</u> are often seen on the <u>bark</u> of trees. Until the 19th century, the only ones found in England were <u>light</u> in colour. Then some areas became <u>polluted</u> and the soot darkened the tree trunks. A <u>black</u> variety of moth was found. The moths had <u>adapted</u> to stay <u>camouflaged</u>.

*2) **Bacteria** adapt to beat antibiotics*

The "<u>survival of the fittest</u>" (see next page) affects bacteria just the same as other living things. They adapt to become <u>resistant</u> to our bacteria-fighting weapons — <u>antibiotics</u>.

1) If someone gets ill they might be given an <u>antibiotic</u> which <u>kills</u> 99% of the bacteria.

2) The 1% that survive are <u>resistant</u> so if they're passed on to somebody else, the antibiotic won't help.

Nowadays bacteria are getting resistant at such a rate the development of antibiotics <u>can't keep up</u>.

*3) **Rats** adapt to beat poison*

The poison <u>warfarin</u> was widely used to control the <u>rat</u> population. However, a certain gene gives rats <u>resistance</u> to it, so rats which carry it are more likely to survive and breed. This gene has become more and more <u>frequent</u> in the rat population, so warfarin isn't as much use any more.

*Environmental change can cause **extinction***

1) The <u>dinosaurs</u> and <u>woolly mammoths</u> became <u>extinct</u>, and it's only <u>fossils</u> that tell us they ever existed.

There are <u>three ways</u> a species can become <u>extinct</u>:

1. The <u>environment changes</u> more quickly than the species can adapt.

2. A new <u>predator</u> or <u>disease</u> kills them all.

3. They can't <u>compete</u> with another (new) species for <u>food</u>.

2) As the environment changes, it'll <u>gradually</u> favour certain characteristics (see p93).

3) Over many generations those features will be present in <u>more</u> of the population. In this way, the species constantly <u>adapts</u> to its changing environment.

4) But if the environment changes too <u>fast</u> the whole species may become <u>extinct</u>.

Natural Selection

OK... Charles Darwin realised that evolution took place, and developed a theory about how it actually happened. He called it the theory of natural selection. This is how he came up with it...

Darwin made **four** important **observations**...

1) All organisms produce more offspring than could possibly survive (e.g. only a few frogspawn survive and become frogs).

2) But in fact, population numbers tend to remain fairly constant over long periods of time.

3) Also, organisms in a species show wide variation in characteristics.

4) Some of the variations are inherited, and so passed on to the next generation.

...and then made these **two deductions**:

1) Since most offspring don't survive, all organisms must have to struggle for survival. Being eaten, disease and competition cause large numbers of individuals to die.

2) The ones who have characteristics that allow them to survive and reproduce better (i.e. the most useful adaptations to the environment) will pass on these characteristics.

This is the famous "survival of the fittest" statement. Organisms with slightly less survival value will probably perish first, leaving the fittest to pass on their genes to the next generation.

Natural selection... sounds like vegan chocolates

This is a good example of how scientific theories come about — someone observes something and tries to explain it. Their theory will then be tested by other scientists using evidence — if the theory passes these tests, it gains in credibility. If not, it's rejected. Natural selection hasn't been rejected yet.

Natural Selection

Have a look at this example to see how a <u>whole population</u> can change its <u>characteristics</u> pretty easily.

*Organisms with certain **characteristics** will **survive** better*

Here's an example...

Once upon a time maybe all rabbits had <u>short ears</u> and managed OK.

Then one day out popped a rabbit with <u>big ears</u> who could hear better and was always the first to dive for cover at the sound of a predator.

Pretty soon he's fathered a whole family of rabbits with <u>big ears</u>, all diving for cover before the other rabbits.

Before you know it there are only <u>big-eared</u> rabbits left — because the rest just didn't hear trouble coming quick enough.

This is how populations <u>adapt</u> to changes in their environment (an organism doesn't actually change when it's alive — changes only occur from generation to generation).
Over many generations the <u>characteristic</u> that <u>increased survival</u> becomes <u>more common</u> in the population. If members of a species are separated somehow, and evolve in different ways to adapt to different conditions, then over time, you can end up with totally <u>different species</u>.

*Darwin's theory **wasn't popular** at first*

1) Darwin's theory <u>caused some trouble</u> at the time — it was the first plausible explanation for our own existence <u>without</u> the need for a "Creator".

2) This was <u>bad news</u> for the religious authorities of the time, who tried to ridicule old Charlie's ideas. The idea that humans and monkeys had a common ancestor was hard for people to accept, and easy to take the mick out of. But, as they say, "<u>the truth will out</u>".

3) Some <u>scientists weren't keen</u> either, at first. Darwin didn't provide a proper explanation of exactly <u>how</u> individual organisms passed on their survival characteristics to their offspring.

4) Later, the idea of <u>genetics</u> was understood — which <u>did</u> explain how characteristics are inherited.

You'll increase your chances of survival by learning this stuff

Darwin was ridiculed by the Church about his theory, but it wasn't the first time a scientist or philosopher was picked on by the Church... Galileo was put under house arrest by the Church for nine years for supporting Copernicus' theory that the Earth was not the centre of the Universe.

Adaptation

Animals and plants survive in many different <u>environments</u> — from <u>hot deserts</u> to <u>cold polar regions</u>, and just about everywhere in between. They can do this because they've <u>adapted</u> to their environment.

Desert animals *have adapted to* save water

Animals that live in <u>hot</u>, <u>dry</u> conditions need to <u>keep cool</u> and use <u>water</u> efficiently.

Large surface area *compared to* volume

This lets desert animals <u>lose more body heat</u> — which helps to stop them overheating.

Efficient with water

1) Desert animals <u>lose less water</u> by producing small amounts of <u>concentrated urine</u>.

2) They also make very little <u>sweat</u>. Camels are able to do this by tolerating <u>big changes</u> in <u>body temperature</u>, while kangaroo rats live in <u>burrows</u> underground where it's <u>cool</u>.

Good in hot, sandy *conditions*

1) Desert animals have very thin layers of <u>body fat</u> to help them <u>lose</u> body heat. Camels keep nearly all their fat in their <u>humps</u>.

2) <u>Large feet</u> spread their <u>weight</u> across soft sand — making getting about easier.

3) A <u>sandy colour</u> gives good <u>camouflage</u> — so they're not as easy for their <u>predators</u> to spot.

Arctic animals *have adapted to* reduce heat loss

Small surface area *compared to* volume

Animals that live in <u>really cold</u> conditions need to <u>keep warm</u>.

Well insulated

Animals living in <u>cold</u> conditions have a <u>compact</u> (rounded) shape to keep their <u>surface area</u> to a minimum — this <u>reduces heat loss</u>.

1) They also have a thick layer of <u>blubber</u> for <u>insulation</u> — this also acts as an <u>energy store</u> when food is scarce.

2) <u>Thick hairy coats</u> keep body heat in, and <u>greasy fur</u> sheds water (this <u>prevents cooling</u> due to evaporation).

Good in snowy *conditions*

1) Arctic animals have <u>white fur</u> to match their surroundings — for <u>camouflage</u>.

2) <u>Big feet</u> help by <u>spreading weight</u> — which stops animals sinking into the snow or breaking thin ice.

Adaptation

Whether you're an animal or a plant, you have to adapt to your environment — and that includes adapting to deal with other plants and animals...

Some **plants** have adapted to living in a **desert**

Desert-dwelling plants make best use of what little water is available.

Minimising water loss

1) Cacti have <u>spines instead of leaves</u> — to <u>reduce water loss</u>.
2) They also have a <u>small surface area</u> compared to their size (about 1000 times smaller than normal plants), which also <u>reduces water loss</u>.
3) A cactus <u>stores water</u> in its thick stem.

Maximising water absorption

Some cacti have <u>shallow</u> but <u>extensive roots</u> to <u>absorb</u> water quickly over a large area. Others have <u>deep roots</u> to access <u>underground water</u>.

Some **plants and animals** are adapted to **deter predators**

There are various <u>special features</u> used by animals and plants to help <u>protect</u> them against being <u>eaten</u>.

1) Some plants and animals have <u>armour</u> — like roses (with <u>thorns</u>), cacti (with <u>sharp spines</u>) and tortoises (with <u>hard shells</u>).

2) Others produce <u>poisons</u> — like bees and poison ivy.

3) And some have amazing <u>warning colours</u> to scare off predators — like wasps.

In a nutshell, it's horses for courses

It's <u>no accident</u> that animals and plants look like they do. So by looking at an animal's <u>characteristics</u>, you should be able to have a pretty good guess at the kind of <u>environment</u> it lives in — or vice versa. Why does it have a large/small surface area... what are those spines for... why is it green... and so on.

Warm-Up and Exam Questions

The warm-up questions run quickly over the basic facts you'll need in the exam. The exam questions come later — but unless you've learnt the facts first you'll find the exams tougher than stale bread.

Warm-up Questions

1) What can a fossil of an organism tell us?
2) Give three adaptations that desert animals have, which enable them to survive hot, dry conditions.
3) What is meant by the term 'survival of the fittest'?
4) Explain the difference between evolution and natural selection.

Exam Questions

1   Which of the following could **not** cause a species to become extinct?

   A   Competition from another species for food.

   B   The environment changing more quickly than the species can adapt.

   C   The species producing too many offspring.

   D   A new disease emerging.

(1 mark)

2   Arctic animals are adapted to their environment.
   The table below is about the adaptations of a polar bear.
   Match each adaptation, **A**, **B**, **C** and **D** in the list with its purpose (**1 - 4**).

   A   Large feet

   B   Thick layer of blubber

   C   White fur

   D   Greasy fur

	Purpose
1	Prevents cooling due to evaporation.
2	Stops them sinking into snow.
3	Helps them retain body heat.
4	Makes it hard for prey to spot them.

(4 marks)

3   Charles Darwin developed the theory of evolution by natural selection.

   (a)   Give one observation he made that led him to this theory.

(1 mark)

   (b)   Explain how natural selection occurs.

(4 marks)

   (c)   Why did Darwin have trouble getting his theory accepted?

(1 mark)

Revision Summary for Section Three

There's a lot to remember from this section and quite a few of the topics are controversial, e.g. cloning, genetic engineering, and so on. You need to know all sides of the story, as well as all the facts... so, here are some questions to help you. If you get any wrong, go back and learn the stuff.

1) What are the two types of variation? Describe their relative importance for plants and animals.

2) List four features of animals which aren't affected at all by environment, and three which are.

3) Draw a set of diagrams showing the relationship between: cell, nucleus, chromosomes, genes, DNA. Describe what genes do.

4) How many pairs of chromosomes do human cells have?

5) Give a definition of mitosis. Draw a set of diagrams showing what happens in mitosis.

6) What is asexual reproduction? Give a proper definition for it. How does it involve mitosis?

7) How do gametes form? How many chromosomes does a human gamete have?

8) Draw out a sequence of diagrams showing how chromosomes combine during sexual reproduction.

9) What is a mutation? Name three things that can increase the likelihood of genetic mutations.

10)*Draw a genetic diagram for a cross between a man with blue eyes (bb) and a woman who has green eyes (Bb). The gene for blue eyes (b) is recessive.

11) Describe two ways in which a faulty gene could lead to health problems.

12) What's the basic idea behind gene therapy?

13) Give a good account of the important stages of genetic engineering.

14) State two examples of useful applications of genetic engineering.

15) Why are some people concerned about genetic engineering?

16) How would you make a plant clone using tissue culture? What are the pros and cons?

17) Explain how adult cell cloning could be used to treat disease. Why do some people think it's wrong?

18) Give three uses of glucose in plants.

19) What are the three limiting factors in photosynthesis?

20) How could you estimate a population size in a habitat using a quadrat? Describe two reasons why your results might not be 100% accurate.

21) Explain two reasons why different species may look similar.

22) Why are pyramids of biomass always pyramid shaped?

23) How much energy and biomass pass from one trophic level to the next? Where does the rest go?

24) Name three things that: a) plants compete for, b) animals compete for.

25) Sketch a graph of prey and predator populations and explain the pattern shown.

26) What is the difference between a parasitic and a mutualistic relationship? Give an example of each.

27) What is a fossil? Describe the three ways that fossils can form. Give an example of each type.

28) Why are fossils important to our understanding of dinosaurs?

29) Why is the fossil record incomplete?

30) What's the theory of evolution?

31) Describe three examples of evolution that we have noticed happening recently.

32) Give three reasons why some species become extinct.

33) What were Darwin's four observations and two deductions? Why was his theory controversial?

34) Explain how an animal that lives in the Arctic might be adapted to its environment.

35) State three ways that plants and animals might be adapted to deter predators.

* Answers on page 290.

Atoms

Hello, good evening and welcome to Chemistry. This section covers all of Chemistry's essential <u>gory details</u> — about <u>atoms</u>, <u>their innards</u>, and <u>what they get up to</u> with each other when no one's looking.

Structure of the *atom* — *there's nothing to it*

The structure of atoms is quite simple. Just learn and enjoy, my friend.

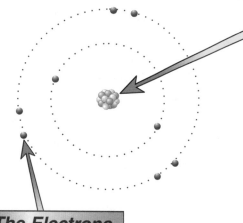

The Nucleus

1) It's in the <u>middle</u> of the atom. It contains <u>protons</u> and <u>neutrons</u>. (It's the <u>number of protons</u> in an atom that decides what element it is.)

2) The nucleus has an overall <u>positive charge</u> because protons are positively charged while neutrons have no charge.

3) Almost the <u>whole mass</u> of the atom is <u>concentrated</u> in the <u>nucleus</u>. But size-wise it's <u>tiny</u> compared to the atom as a whole.

The Electrons

1) They move <u>around</u> the nucleus in energy levels called <u>shells</u>. (Each shell is only allowed a <u>certain number of electrons</u>.)

2) They have a <u>negative charge</u> (electrons and protons have equal but opposite charges).

3) They're <u>tiny</u> compared to the nucleus (they have <u>virtually no mass</u>), but as they move around they cover a <u>lot of space</u>. (The size of their orbits determines <u>how big</u> the atom is.)

Number of protons equals *number of electrons*

1) Neutral atoms have <u>no charge</u> overall.

2) This is because the <u>number of protons</u> always <u>equals</u> the <u>number of electrons</u> in a <u>neutral atom</u>, and the <u>charge</u> on the <u>electrons</u> is the <u>same</u> size as the charge on the <u>protons</u>, but <u>opposite</u>.

3) The number of neutrons isn't fixed but is usually <u>about the same</u> as the number of protons.

Know your *particles*

1) <u>Protons</u> are <u>heavy</u> and <u>positively charged</u>.

2) <u>Neutrons</u> are <u>heavy</u> and <u>neutral</u> (no charge).

3) <u>Electrons</u> are <u>tiny</u> and <u>negatively charged</u>.

Particle	Relative mass	Relative charge
Proton	1	+1
Neutron	1	0
Electron	$\frac{1}{2000}$	-1

Each element has an atomic number and a mass number

1) The <u>atomic number</u> says how many <u>protons</u> there are in an atom, and is unique to that element.

2) The atomic number <u>also</u> tells you the number of <u>electrons</u>.

3) The <u>mass number</u> is the total number of <u>protons and neutrons</u> in the atom. So if you want to find the number of <u>neutrons</u> in an atom, just <u>subtract</u> the <u>atomic number</u> from the <u>mass number</u>.

MASS NUMBER
Total number of protons and neutrons.

ATOMIC NUMBER (OR PROTON NUMBER)
Number of protons, which is equal to the number of electrons.

Elements, Compounds and Mixtures

There are only about <u>100 or so</u> different kinds of atoms, which doesn't sound too bad.
But they can <u>join together</u> in <u>loads</u> of different combinations, which makes life more complicated.

Elements consist of *one type* of *atom* only

Quite a lot of everyday substances are <u>elements</u>:

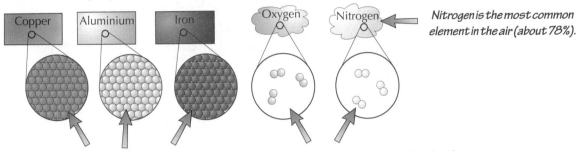

Nitrogen is the most common element in the air (about 78%).

The atoms in solids are tightly packed.

Atoms in gases often go round in pairs. A molecule with two atoms in it is called a <u>diatomic molecule</u>.

Compounds are *chemically bonded*

A <u>compound</u> is a substance that is made of <u>two or more</u>
different <u>elements</u> which are chemically <u>joined</u> (<u>bonded</u>) together.

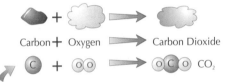

Carbon + Oxygen → Carbon Dioxide

C + OO → OCO CO_2

1) For example, <u>carbon dioxide</u> is a <u>compound</u> formed from a <u>chemical reaction</u>. One carbon atom
reacts with two oxygen atoms to form a <u>molecule</u> of carbon dioxide, with the <u>formula</u> CO_2.

2) It's <u>very difficult</u> to <u>separate</u> the two original elements out again.

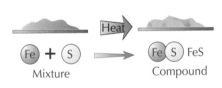

Fe + S → Fe S FeS
Mixture   Compound

3) The <u>properties</u> of a compound are often <u>totally different</u>
from the properties of the <u>original elements</u>.

4) For example, if a mixture of iron and sulphur is <u>heated</u>, the iron
and sulphur atoms react to form the compound <u>iron sulphide</u>
(FeS). Iron sulphide is not much like iron (e.g. it's not attracted
to a magnet), nor is it much like sulphur (e.g. it's not yellow
in colour).

Mixtures are *easily separated* — *not like compounds*

1) Unlike in a compound, there's <u>no chemical bond</u> between the different parts of a mixture.
The parts can be separated out by <u>physical methods</u> such as distillation (see pages 121 and 124).

2) <u>Air</u> is a <u>mixture</u> of gases, mainly nitrogen, oxygen, carbon dioxide and argon.
The gases can all be <u>separated out</u> fairly easily.

3) The <u>properties</u> of a mixture are just a <u>mixture</u> of the properties of the <u>separate parts</u>.

4) A <u>mixture</u> of <u>iron powder</u> and
<u>sulphur powder</u> will show the
properties of <u>both iron and sulphur</u>.
It will contain grey magnetic bits of
iron and bright yellow bits of sulphur.

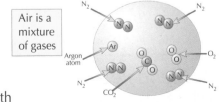

Air is a
mixture
of gases

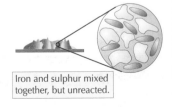

Iron and sulphur mixed
together, but unreacted.

5) <u>Crude oil</u> is a <u>mixture</u> of different length
hydrocarbon molecules — see page 124.

The Periodic Table

In the 1800s chemists were keen to try and find <u>patterns</u> in the elements they knew about. And the <u>more</u> elements that were identified, the <u>clearer</u> those patterns became...

Dmitri **Mendeleev** arranged the **elements** in **groups**

Mendeleev's Table of the Elements

H																	
Li	Be											B	C	N	O	F	
Na	Mg											Al	Si	P	S	Cl	
K	Ca	*	Ti	V	Cr	Mn	Fe	Co	Ni	Cu	Zn	*	*	As	Se	Br	
Rb	Sr	Y	Zr	Nb	Mo	*	Ru	Rh	Pd	Ag	Cd	In	Sn	Sb	Te	I	
Cs	Ba	*	*	Ta	W	*	Os	Ir	Pt	Au	Hg	Tl	Pb	Bi			

1) In <u>1869</u>, a Russian scientist called <u>Dmitri Mendeleev</u> arranged the 50 or so known elements in order of <u>atomic mass</u> to make a Table of Elements.

2) Mendeleev's table placed elements with <u>similar chemical properties</u> in the same <u>vertical groups</u> — but he found that he had to leave <u>gaps</u> in his table to make this work.

3) The <u>gaps</u> in Mendeleev's table of elements were really clever because they <u>predicted</u> the properties of <u>undiscovered elements</u>.

4) Since then <u>new elements</u> have been found which <u>fit into the gaps</u> in Mendeleev's table. Over the last hundred years or so the table has been <u>refined</u> to produce the <u>periodic table</u> we know (and love) today...

Elementary my dear Mendeleev

Even though its not the periodic table we use today, its important to know how much of an influence Mendeleev's periodic table has been on our modern periodic table. Make sure you know how Mendeleev arranged his table and how it came to look like the one we're used to using today.

The Periodic Table

The periodic table is really important. You can try to do chemistry without it, but it's likely to all end in disaster. Learn the rules, know the trends and practise safe chemistry.

The *periodic table* puts *elements* with *similar properties* together

1) The periodic table is laid out so that elements with <u>similar properties</u> form <u>columns</u>.

2) These <u>vertical columns</u> are called <u>groups</u> and Roman numerals are often used for them.

3) If you know the <u>properties</u> of <u>one element</u>, you can <u>predict</u> properties of <u>other elements</u> in that group.

4) For example the <u>Group I</u> elements are Li, Na, K, Rb, Cs and Fr. They're all <u>metals</u> and they <u>react the same way</u>. E.g. they all react with water to form an <u>alkaline solution</u> and <u>hydrogen gas</u>.

5) You can also make predictions about <u>reactivity</u>. E.g. in Group 1, the elements react <u>more vigorously</u> as you go <u>down</u> the group. And in Group 7, <u>reactivity decreases</u> as you go down the group.

6) There are <u>100ish elements</u>, which all materials are made of. If it wasn't for the periodic table <u>organising everything</u>, you'd have a <u>heck of a job</u> remembering all those properties. It's <u>ace</u>.

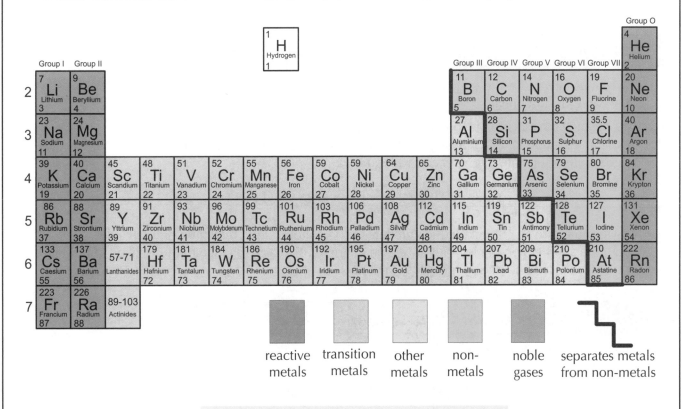

reactive metals   transition metals   other metals   non-metals   noble gases   separates metals from non-metals

The periodic table gives you the mass number and atomic number of all known elements.

mass number (total number of protons & neutrons) ➡ 4 He Helium

atomic number (number of protons) ➡ 2

The periodic table is all you need

One important thing to understand on these pages is that a <u>scientific theory</u> (such as "elements can be grouped in a table according to their properties") can be used to <u>make predictions</u> (such as "there are gaps in the table so there must be some undiscovered elements to fill those gaps"). Got it... good.

Balancing Equations

All chemical reactions can be shown using an _equation_. Unfortunately, getting equations right takes a bit of practice. So make sure you _get_ a bit of practice — don't just skate over them.

Atoms **aren't lost or made** in chemical reactions

1) During chemical reactions, things _don't_ appear out of nowhere and things _don't_ just disappear.

2) You still have the _same atoms_ at the _end_ of a chemical reaction as you had at the _start_. They're just _arranged_ in different ways.

3) _Balanced symbol equations_ show the atoms at the _start_ (the _reactant_ atoms) and the atoms at the _end_ (the _product_ atoms) and how they're arranged. For example:

Word equation: Magnesium + Oxygen → Magnesium oxide

Balanced symbol equation: $2Mg + O_2 → 2MgO$

4) Because atoms aren't gained or lost, the _mass_ of the reactants _equals_ the mass of the products. So, if you react _6 g of magnesium_ with _4 g of oxygen_, you'd end up with _10 g of magnesium oxide_.

Balancing the equation — match them up one by one

1) There must always be the _same_ number of atoms of each element on _both sides_ — they can't just _disappear_.

2) You _balance_ the equation by putting numbers _in front_ of the formulas where needed. Take this equation for reacting sulphuric acid (H_2SO_4) with sodium hydroxide (NaOH) to get sodium sulphate (Na_2SO_4) and water (H_2O):

$$H_2SO_4 + NaOH → Na_2SO_4 + H_2O$$

The _formulas_ are all correct but the numbers of some atoms _don't match up_ on both sides. E.g. there are 3 H's on the left, but only 2 on the right. You _can't change formulas_ like H_2O to H_3O. You can only put numbers _in front of them_:

Method: balance just **ONE** type of atom at a time

The more you practise, the quicker you get, but all you do is this:

1) Find an element that _doesn't balance_ and _pencil in a number_ to try and sort it out.

2) _See where it gets you._ It may create _another imbalance_ — if so, just pencil in _another number_ and see where that gets you.

3) Carry on chasing _unbalanced_ elements and it'll _sort itself out_ pretty quickly.

I'll show you. In the equation above you soon notice we're short of H atoms on the right-hand side.
1) The only thing you can do about that is make it $2H_2O$ instead of just H_2O:

$$H_2SO_4 + NaOH → Na_2SO_4 + 2H_2O$$

2) But that now causes too many H atoms and O atoms on the right-hand side, so to balance that up you could try putting 2NaOH on the left-hand side:

$$H_2SO_4 + 2NaOH → Na_2SO_4 + 2H_2O$$

3) And suddenly there it is! _Everything balances._ And you'll notice the Na just sorted itself out.

Warm-Up and Exam Questions

It's easy to think you've learnt everything in the section until you try the warm-up questions. Don't panic if there are bits you've forgotten. Just go back over those bits until they're firmly fixed in your brain.

Warm-up Questions

1) Give the charges of the three types of particle which make up atoms.
2) In a neutral atom, which particles are always equal in number?
3) Explain the difference between mass number and atomic number.
4) Name an element in which the atoms are tightly packed at room temperature.
5) Balance this equation for the reaction of glucose ($C_6H_{12}O_6$) and oxygen:
$$C_6H_{12}O_6 \ + \ O_2 \ \rightarrow \ CO_2 \ + \ H_2O$$
6) Draw and label an atom of nitrogen.
7) In this equation: $H_2SO_4 + 2NaOH \rightarrow Na_2SO_4 + 2H_2O$
explain the difference between the meaning of the $_2$ in H_2SO_4 and the 2 in 2NaOH.

Exam Questions

1 The proton has a relative mass of 1. What is the relative mass of the neutron?

A 2000

B 1

C 1/2000

D 2

(1 mark)

2 Which of the following statements about Mendeleev's periodic table is **not** true?

A Elements with similar chemical properties were placed in vertical groups.

B Gaps were left which helped in predicting the properties of undiscovered elements.

C The elements were arranged in order of atomic mass.

D Mendeleev's periodic table contained over 100 elements.

(1 mark)

3 The modern periodic table can be divided into metals and non-metals. The non-metals are

A on the left of the periodic table

B on the right of the periodic table

C in the middle of the periodic table

D in Group II

(1 mark)

Exam Questions

4   Air is a mixture of gases, mainly nitrogen, oxygen, carbon dioxide and argon.

   (a) How is a compound different from a mixture?

   (1 mark)

   (b) (i)   What group is argon in?

   (1 mark)

      (ii) Give the mass number of argon.

   (1 mark)

   (c) Name one gas found in air which is a compound.

   (1 mark)

   (d) Oxygen gas, O_2, reacts with magnesium to form magnesium oxide, MgO.
        Write a balanced symbol equation for this reaction.

   (2 marks)

5   Which of these statements about chemical reactions is **not** true?

   A   The mass of the reactants is always equal to the mass of the products.

   B   Atoms are neither created nor destroyed in a reaction.

   C   The mass of the products is always less than the mass of the reactants.

   D   In a written equation, the mass of all the atoms on the left of the arrow
        is equal to the mass of all the atoms on the right of the arrow.

   (1 mark)

6   Sulphuric acid, H_2SO_4, reacts with ammonia, NH_3, to form ammonium sulphate,
   $(NH_4)_2SO_4$.

   (a) Write the word equation for this reaction.

   (1 mark)

   (b) Write a balanced symbol equation for this reaction.

   (2 marks)

   (c) In the balanced equation, how many atoms are there in the reactants?

   (1 mark)

Group 1 — The Alkali Metals

Alkali metals all have <u>one electron</u> in their outer shell. The atoms are keen to <u>get rid</u> of this sole lingering electron — which makes them very <u>reactive</u>. <u>Lithium</u>, <u>sodium</u> and <u>potassium</u> are stars of the show here.

Group 1 elements react vigorously in water

1) When <u>lithium</u>, <u>sodium</u> or <u>potassium</u> are put in <u>water</u>, they react <u>vigorously</u>.

2) The <u>reaction</u> makes an <u>alkaline</u> solution (which would change <u>universal indicator</u> to <u>blue</u> or <u>purple</u>) — this is why Group 1 is known as the <u>alkali metals</u>.

3) As you go <u>down</u> Group 1, the <u>lingering electron</u> is in a shell that's <u>further from the nucleus</u>. This means the electron is <u>easier to get rid of</u>. That makes the elements further down Group 1 <u>more reactive</u>.

4) You can see this in the <u>rate of reaction</u> with water (i.e. the time taken for a lump of the same size of each element to <u>react completely</u> with the water and disappear). <u>Lithium</u> takes longer than sodium or potassium to react, so it's the <u>least reactive</u>. <u>Potassium</u> takes the shortest time to react of these three elements, so it's the <u>most reactive</u>.

> The elements in <u>GROUP 1</u> get <u>MORE REACTIVE</u> as the <u>ATOMIC NUMBER INCREASES</u>.

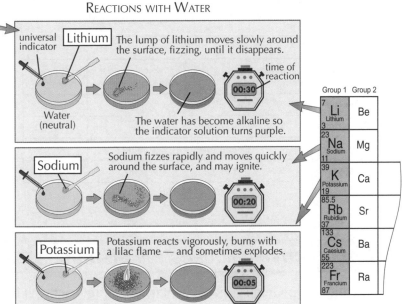

REACTIONS WITH WATER

Lithium The lump of lithium moves slowly around the surface, fizzing, until it disappears.
The water has become alkaline so the indicator solution turns purple.

Sodium Sodium fizzes rapidly and moves quickly around the surface, and may ignite.

Potassium Potassium reacts vigorously, burns with a lilac flame — and sometimes explodes.

Reaction with water produces hydrogen gas

1) The <u>reaction</u> of the alkali metals with water produces <u>hydrogen</u> — this is what you can see <u>fizzing</u>.

2) A <u>lighted splint</u> will <u>indicate</u> the hydrogen by making the notorious "<u>squeaky pop</u>" as the H_2 ignites.

3) These reactions can be written down as <u>chemical equations</u> — e.g. for <u>sodium</u> the equation is...

> In words: sodium + water $\rightarrow$ sodium hydroxide + hydrogen
> In symbols: $2Na(s) + 2H_2O(l) \rightarrow 2NaOH(aq) + H_2(g)$

<u>STATE SYMBOLS</u>: (s) = <u>solid</u>, (l) = <u>liquid</u>, (aq) = <u>aqueous</u> (dissolved in water), (g) = <u>gas</u>

Alkali metals like to bond with other atoms and molecules

1) Alkali metals only have <u>one outer electron</u>, so they're <u>really keen</u> to <u>get rid</u> of it and get back to a <u>full outer shell</u>.

2) Getting a full outer shell is an atom's main aim in life — this is <u>why</u> alkali metals are so <u>happy to react</u> and <u>form compounds</u> with other elements.

Group 7 — The Halogens

The halogens are all <u>one electron short</u> of having a full outer shell.

HALOGEN — seven letters — Group 7

1) The elements in <u>Group 7</u> of the periodic table are called the <u>halogens</u>.

2) Halogens are <u>really useful</u> and many <u>everyday things</u> contain them, e.g. <u>toothpaste</u> and the <u>non-stick coating</u> on frying pans. But on their own, they're <u>poisonous</u>.

3) <u>Chlorine's</u> good because it <u>kills bacteria</u>. It's used in <u>bleach</u> and <u>swimming pools</u>.

4) The <u>properties</u> of the elements in <u>Group 7</u> change <u>gradually</u> as you go <u>down</u> the group (i.e. as the atomic number <u>increases</u>). Look at the table below.

Group 7 Elements	Properties			
	Atomic number	Colour	Physical state at room temperature	Boiling point
Fluorine	9	yellow	gas	−188 °C
Chlorine	17	green	gas	−34 °C
Bromine	35	red-brown	liquid	59 °C
Iodine	53	dark grey	solid	185 °C

5) As the <u>atomic number</u> of the halogens <u>increases</u>, the elements have a <u>darker colour</u> and a <u>higher boiling point</u> (which is why they go from <u>gases</u> at the top of Group 7 to <u>solids</u> at the bottom, at room temperature).

More reactive halogens will displace less reactive ones

1) The <u>higher up</u> Group 7 an element is, the <u>more reactive</u> it is. This is because the shell with the missing electron is <u>nearer to the nucleus</u>, so the pull from the <u>positive nucleus</u> is <u>greater</u>.

2) A <u>displacement reaction</u> is where a <u>more reactive</u> element "<u>pushes out</u>" (displaces) a <u>less reactive</u> element from a compound.

3) For example, <u>chlorine</u> is more reactive than <u>iodine</u> (it's higher up Group 7). Chlorine therefore reacts with potassium iodide to form <u>potassium chloride</u>, and the iodine gets left in the solution.

$$Cl_2(g) + 2KI(aq) \rightarrow I_2(aq) + 2KCl(aq)$$
$$Cl_2(g) + 2KBr(aq) \rightarrow Br_2(aq) + 2KCl(aq)$$

These are the equations for chlorine displacing iodine and bromine. They might give you a different example in the exam, but the equations are all quite similar.

Halogen atoms like to <u>pair up</u> to form <u>diatomic molecules</u>.

The halogens react with metals to form salts

Halogens react with most <u>metals</u>, including <u>iron</u> and <u>aluminium</u>, to form <u>salts</u> (also called <u>metal halides</u>).

$$2Al(s) + 3Cl_2(g) \rightarrow 2AlCl_3(s) \quad \text{(Aluminium chloride)}$$
$$2Fe(s) + 3Br_2(g) \rightarrow 2FeBr_3(s) \quad \text{(Iron(III) bromide)}$$

The aluminium and iron ions both have a 3+ charge. That means they need three halogen ions (1-) for the charges to be balanced. This is why these two formulae are $AlCl_3$ and $FeBr_3$.

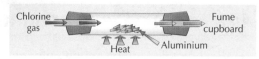

Group 0 — The Noble Gases

The noble gases — stuffed full of every honourable virtue.

Group 0 elements are all inert, colourless gases

1) Group 0 elements are called the <u>noble gases</u> and include the elements <u>helium</u>, <u>neon</u> and <u>argon</u> (plus a few others).

2) The noble gases were only <u>discovered</u> just over 100 years ago — it took so long to find them because they have properties that make them <u>hard to observe</u>...

3) All elements in Group 0 are <u>colourless gases</u> at room temperature.

4) They are also more or less <u>inert</u> — this means they <u>don't react</u> with much at all. (The reason for this is that they have a <u>full outer shell</u>. This means they're <u>not</u> desperate to <u>give up</u> or <u>gain</u> electrons.)

5) Luckily the noble gases all have a dead handy property that lets you <u>see</u> them — they each give out light if you pass an <u>electric current</u> through them. Each noble gas gives out a particular colour of light.

		Group O
		4 **He** Helium 2
Group 6	Group 7	20 **Ne** Neon 10
O	F	
S	Cl	40 **Ar** Argon 18
Se	Br	84 **Kr** Krypton 36
Te	I	131 **Xe** Xenon 54
Po	At	222 **Rn** Radon 86

The noble gases have many everyday uses...

Neon is used in electrical discharge tubes

<u>Neon lights</u> are used in tacky <u>shop signs</u> — the kind you'd expect to see if you visited Las Vegas. They don't use much <u>current</u> so they're cheap to run, and they give out a <u>bright red light</u>.

Noble gases are used in lasers too

There's the famous red <u>helium-neon</u> laser and the more powerful <u>argon laser</u>.

Helium is used in airships and party balloons

Helium has a <u>lower density</u> than air — so it makes balloons <u>float</u>. And it's a lot safer to use than hydrogen (the famous airship Hindenburg was filled with hydrogen and caught <u>fire</u>).

Argon is used in filament lamps (light bulbs)

It provides an <u>inert atmosphere</u> which stops the very hot filament from <u>burning away</u>.

Noble, but lazy

Well, they don't react so there's a bit less to learn about the noble gases. Nevertheless, there's likely to be a question or two on them so <u>make sure you learn everything on this page</u>...

Warm-Up and Exam Questions

These questions are all about the groups of the periodic table that you need to know about. Treat the exam questions like the real thing — don't look back through the book until you've finished.

Warm-up Questions

1) In Group 1, as you go down the periodic table, does the reactivity increase or decrease?
2) In Group 7, what is the trend in physical state as you go down the group?
3) Name two noble gases and give a use for each one you have named.
4) Which gas is produced when an alkali metal reacts with water?
5) Give an example of a salt produced when a metal reacts with a Group 7 element.

Exam Questions

1   (a) The noble gases are inert. Explain what this means.

(1 mark)

(b) Explain why the noble gases are inert.

(1 mark)

(c) Which noble gas is used in filament bulbs, and why is it used in this way?

(3 marks)

2   The table shows some of the physical properties of four of the halogens.

Halogen	Properties			
	Atomic number	Colour	Physical state at room temperature	Boiling point
Fluorine	9	yellow		−188 °C
Chlorine	17	green		−34 °C
Bromine	35	red-brown		59 °C
Iodine	53	dark grey		185 °C

(a) Fill in the physical state at room temperature of all four halogens.

(4 marks)

(b) Draw an arrow next to the left hand side of the table to show the direction of increasing reactivity in the halogens.

(1 mark)

(c) This equation shows a reaction between chlorine and potassium iodide.

$$Cl_{2\,(g)} + 2KI_{(aq)} \rightarrow I_{2\,(aq)} + 2KCl_{(aq)}$$

(i) What type of reaction is this?

(1 mark)

(ii) Which is the less reactive halogen in this reaction?

(1 mark)

Properties of Metals

Metals are all similar but slightly different. They have some basic properties in common, but each has its own specific combination of properties, which means you use different ones for different purposes.

*Metals are on the **left** and **middle** of the **periodic table***

Most of the elements are metals — so they cover most of the periodic table.
In fact, only the elements on the far right are non-metals.
The so-called transition metals are found in the centre block of the periodic table.
Many of the metals in everyday use are transition metals — such as titanium, iron and nickel.

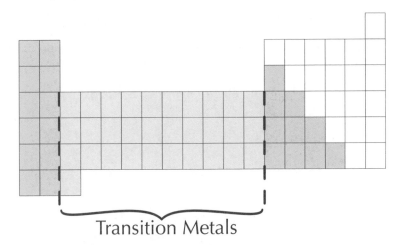

Transition Metals

*Metals are **strong** and **bendy**, and they're **great conductors***

All metals have some fairly similar basic properties.

1) Metals are strong (hard to break), but they can be bent or hammered into different shapes.

2) They're great at conducting heat.

3) They conduct electricity well.

Metals (and especially transition metals) have loads of everyday uses because of these properties...

- Their strength and 'bendability' makes them handy for making into things like bridges and car bodies.

- Metals are ideal if you want to make something that heat needs to travel through, like a saucepan base (see page 198).

- And their conductivity makes them great for making things like electrical wires.

*Transition metals have loads of everyday uses — partly
because they're not crazily reactive like, say, potassium
(which would catch fire if it got rained on).*

Properties of Metals

The reason all metals have the same basic properties is because of the bonding in metals.
It's their exact properties which are used to match metals to their uses.

It's the **structure of metals** that gives them their **properties**

1) <u>All</u> metals have the <u>same</u> basic properties. These are due to the <u>special type of bonding</u> in metals.

2) Metals consist of a <u>giant structure</u> of atoms held together with <u>metallic bonds</u>.

3) These special bonds allow the <u>outer electron(s)</u> of each atom to <u>move freely</u>.

4) This creates a "sea" of <u>free electrons</u> throughout the metal, which is what gives rise to
many of the properties of metals.

5) This includes their <u>conduction</u> of <u>heat</u> and <u>electricity</u> (see page 198).

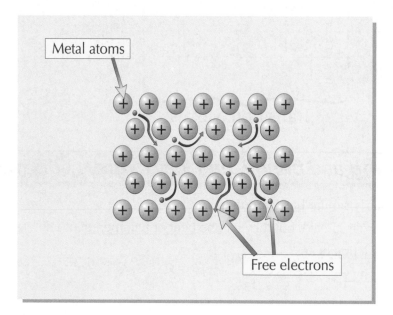

Metal atoms

Free electrons

A metal's **exact properties** decide how it's best **used**

1) The properties above are <u>typical properties</u> of metals.
Not all metals are the same though — their <u>exact properties</u> determine how they're used.

2) If you wanted to make an <u>aeroplane</u>, you'd probably use metal as it's <u>strong</u> and can be
<u>bent into shape</u>, but you'd also need it to be <u>light</u> — so <u>aluminium</u> would be a good choice.

3) And if you were making <u>replacement hips</u>, you'd pick a metal that <u>won't corrode</u> when it
comes in contact with water — it'd also have to be <u>light</u> too, and not too bendy. <u>Titanium</u> has
all of these properties so it's used for this.

Properties of metals are all due to the "sea" of free electrons

So, all metals <u>conduct electricity and heat</u> and can be <u>bent into shape</u>. But lots of them have <u>special</u>
<u>properties</u> too. You have to decide what properties you need and use a metal with those properties.

Extraction of Metals

You don't tend to find big lumps of pure metal in the ground* — the metal atoms tend to be joined to other atoms in compounds. It can be a bit of a tricky, expensive process to separate the metal out. (*You only find <u>unreactive</u> metals, e.g. gold, in their unreacted state. That makes sense if you think about it.)

Ores contain enough metal to make extraction worthwhile

1) <u>Rocks</u> are made of <u>minerals</u>. (Minerals are just <u>solid elements and compounds</u>.)

2) A <u>metal ore</u> is a <u>mineral</u> which contains <u>enough metal</u> to make it <u>worthwhile</u> extracting the metal from it. Ores are "finite resources" — there's only so much of them.

As <u>technology</u> improves, it becomes possible to extract <u>more metal</u> from a sample of rock than previously. So it might now be <u>worth</u> extracting metal that wasn't worth extracting in the past.

3) In many cases the ore is an <u>oxide</u> of the metal. Here are a few examples:

a) A type of <u>iron ore</u> is called <u>haematite</u>. This is iron(III) oxide (Fe_2O_3).

b) The main <u>aluminium ore</u> is called <u>bauxite</u>. This is aluminium oxide (Al_2O_3).

c) A type of <u>copper ore</u> is called <u>chalcopyrite</u>. This is copper iron sulphide ($CuFeS_2$).

Chalcopyrite — a copper ore

Some metals can be extracted by reduction with carbon

1) Electrolysis (splitting with electricity — see page 113) is one way of <u>extracting a metal</u> from its ore. The other common way is chemical <u>reduction</u> using <u>carbon</u> or <u>carbon monoxide</u>.

2) When an ore is <u>reduced</u>, <u>oxygen is removed</u> from it, e.g.

$$Fe_2O_3 \quad + \quad 3CO \quad \rightarrow \quad 2Fe \quad + \quad 3CO_2$$

iron(III) oxide + carbon monoxide $\rightarrow$ iron + carbon dioxide

Extraction of Metals

Its position in the reactivity series determines how it's extracted

1) Metals <u>higher than carbon</u> in the reactivity series have to be extracted using <u>electrolysis</u>, which is expensive.

2) Metals <u>below carbon</u> in the reactivity series can be extracted by <u>reduction</u> using <u>carbon</u>.

   This is because carbon <u>can only take the oxygen</u> away from metals which are <u>less reactive</u> than carbon <u>itself</u> is. For example, <u>iron oxide</u> is reduced in a <u>blast furnace</u> to make <u>iron</u>.

	The Reactivity Series		
Extracted using electrolysis	Potassium	K	more reactive
	Sodium	Na	
	Calcium	Ca	
	Magnesium	Mg	
	Aluminium	Al	
	<u>CARBON</u>	<u>C</u>	
Extracted by reduction using carbon	Zinc	Zn	
	Iron	Fe	
	Tin	Sn	less reactive
	Copper	Cu	

A more reactive metal displaces a less reactive metal

1) <u>More reactive</u> metals react <u>more strongly</u> than <u>less reactive</u> metals — so a metal can be extracted from its <u>oxide</u> by <u>any</u> more reactive metal. E.g. <u>tin</u> could be extracted from <u>tin oxide</u> by more reactive <u>iron</u>.

$$\text{tin oxide} + \text{iron} \rightarrow \text{iron oxide} + \text{tin}$$

2) And if you put a <u>more reactive metal</u> into the solution of a <u>dissolved metal compound</u>, the more reactive metal will <u>replace</u> the <u>less reactive metal</u> in the compound.

$$\text{tin sulphate} + \text{iron} \rightarrow \text{iron sulphate} + \text{tin}$$

3) But if a piece of <u>copper metal</u> is put into a solution of tin sulphate, <u>nothing happens</u>. The more reactive metal (tin) is <u>already</u> in the solution.

Learn how metals are extracted — ore else

Extracting metals isn't cheap. You have to pay for special equipment, energy and labour. Then there's the cost of getting the ore to the extraction plant. If there's a choice of extraction methods, a company always picks the <u>cheapest</u>, unless there's a good reason not to — they're <u>not</u> extracting it for fun.

Extracting Pure Copper

Copper is often dug out of the ground as an ore called malachite. In theory, the copper could be extracted by reducing it with carbon. The problem is that the copper produced by reduction isn't pure enough for use in electrical conductors.

Electrolysis is used to obtain very pure copper

1) Electrolysis means "splitting up with electricity".

2) It requires a liquid (called the electrolyte) which will conduct electricity. Electrolytes are usually free ions dissolved in water. Copper(II) sulphate solution is the electrolyte used in purifying copper — it contains Cu^{2+} ions.

3) The electrical supply acts like an electron pump:

> 1) It pulls electrons off copper atoms at the anode, causing them to go into solution as Cu^{2+} ions.
>
> 2) It then offers electrons at the cathode to nearby Cu^{2+} ions to turn them back into copper atoms.
>
> 3) The impurities are dropped at the anode as a sludge, whilst pure copper atoms bond to the cathode.
>
> 4) The electrolysis can go on for weeks and the cathode is often twenty times bigger at the end of it.

The cathode is the negative electrode. It starts as a thin piece of pure copper and more pure copper adds to it.

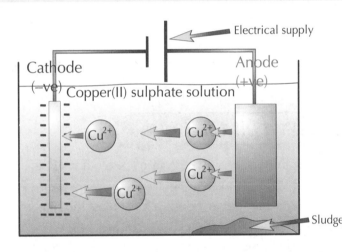

The anode is the positive electrode. It's just a big lump of impure (boulder) copper, which will dissolve.

| Pure copper is deposited on the pure cathode (–ve) | Copper dissolves from the impure anode (+ve) |

The reaction at the cathode is:
$$Cu^{2+}_{(aq)} + 2e^- \rightarrow Cu_{(s)}$$

The reaction at the anode is:
$$Cu_{(s)} \rightarrow Cu^{2+}_{(aq)} + 2e^-$$

Recycling copper saves money and resources

1) The supply of copper-rich ores is limited, so it's important to recycle as much copper as possible.

2) Recycling copper uses only 15% of the energy that'd be used to mine and extract the same amount. Recycling copper helps to conserve fossil fuels and reduce carbon dioxide emissions.

3) Scientists are looking into new ways of extracting copper from low-grade ores (ores that only contain small amounts of copper) or from the waste that is currently produced when copper is extracted.

Electrolysis can take an... awfully... long... time...

Electrolysis ain't cheap — it takes a lot of electricity, which costs money. It's the only way of getting pure enough copper for electrical wires though, so it's worth it.

Other Metals

There are loads of metals. But if none of them have quite the properties you need,
you could try an alloy.

Aluminium is useful, but expensive to extract

1) Aluminium has a low density
   and is corrosion-resistant.

   The aluminium reacts with oxygen in the air to form aluminium oxide. This sticks firmly to the aluminium below and stops any further reaction taking place.

2) These properties make aluminium
   a very useful structural material.
   It can be used for loads of things
   from window frames to electricity
   cables and aircraft.

3) You can't extract aluminium from
   its oxide by the cheap method of
   reduction with carbon. It has
   to be extracted by electrolysis.
   This requires lots of energy, which
   makes it an expensive process.

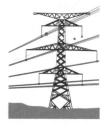

Pure iron tends to be a bit too bendy

1) 'Iron' straight from the blast furnace is only 96% iron. The other 4% is impurities.

2) This impure iron is brittle. It's used for ornamental railings but it doesn't have many
   other uses. So all the impurities are removed from most blast furnace iron.

3) This pure iron has a regular arrangement of identical atoms. The layers of atoms
   can slide over each other, which makes the iron soft and easily shaped. This iron is
   far too bendy for most uses.

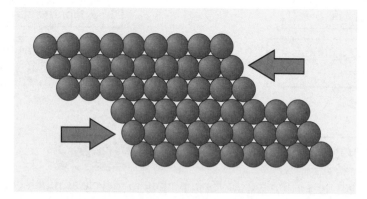

4) Most of the pure iron is changed into alloys called steels. Steels are made
   by adding small amounts of carbon (plus maybe other metals) to the iron.

   An alloy is a mixture of two or more metals, or a mixture of a metal and a non-metal.

Most iron is changed into steel, otherwise it's too bendy or too brittle

The Eiffel Tower is made of iron — but the problem with iron is that it goes rusty if air and water get
to it. So the Eiffel Tower has to be painted every seven years to make sure that it doesn't rust. This is
quite a job and takes an entire year for a team of 25 painters. Too bad they didn't use stainless steel.

Other Metals

Alloys are harder than *pure metals*

1) Different elements have <u>different sized atoms</u>. So when an element such as carbon is added to pure iron, the <u>smaller</u> carbon atom will <u>upset</u> the layers of pure iron atoms, making it more difficult for them to slide over each other. So alloys are <u>harder</u>.

2) Many metals in use today are actually <u>alloys</u>. For example:

> **BRONZE = COPPER + TIN**
>
> Bronze is <u>harder</u> than copper.
> It's good for making medals and statues from.

> **CUPRONICKEL = COPPER + NICKEL**
>
> This is <u>hard</u> and <u>corrosion resistant</u>.
> It's used to make "silver" coins.

> **GOLD ALLOYS ARE USED TO MAKE JEWELLERY**
>
> Pure gold is <u>too soft</u>. Metals such as zinc, copper, silver, palladium and nickel are used to harden the "gold".

> **ALUMINIUM ALLOYS ARE USED TO MAKE AIRCRAFT**
>
> Aluminium has a <u>low density</u>, but it's <u>alloyed</u> with small amounts of other metals to make it <u>stronger</u>.

3) In the past, the development of alloys was by <u>trial and error</u>. But nowadays we understand much more about the properties of metals, so alloys can be <u>designed</u> for specific uses.

Smart alloys return to their *original shape*

1) <u>Nitinol</u> is a "<u>shape memory alloy</u>" — it has a <u>shape memory</u> property.

2) If you <u>bend</u> a wire made of this smart alloy, it'll go back to its <u>original shape</u> when it's <u>heated</u>. You can get specs with frames made from a smart alloy — you can sit on them and not destroy them.

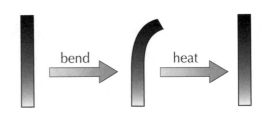

3) At the moment, <u>metal fatigue</u> in smart alloys is a lot <u>worse</u> than in normal alloys. Smart alloys are also <u>more expensive</u> than steel or aluminium.

Alloys are really important in industry

If the properties of a metal aren't quite suited to a job, an alloy is often used instead. To make an alloy you mix one metal with another metal or non-metal. The finished alloy can be a lot harder, or less brittle — the properties can be varied and they can be made to suit a particular job really well.

Warm-Up and Exam Question

The warm-up questions run quickly over the basic facts you'll need in the exam.
Unless you've learnt the facts first you'll find the exam questions pretty difficult.

Warm-up Questions

1) Name two transition metals.
2) Give three useful physical properties of most metals.
3) What type of bonding do metals contain?
4) What is the name for a rock which contains a metal chemically bonded to other elements?
5) Name a metal which can be extracted from its ore by reduction with carbon.
6) What is an alloy? Give two examples, with a use for each.

Exam Questions

1   The diagram shows part of the reactivity series of metals, together with carbon.

Potassium	K	*more*
Sodium	Na	*reactive*
Calcium	Ca	
Magnesium	Mg	
Aluminium	Al	
<u>CARBON</u>	<u>C</u>	
Zinc	Zn	
Iron	Fe	
Tin	Sn	*less*
Copper	Cu	*reactive*

(a) Name one metal which is extracted from its ore using electrolysis.

(1 mark)

(b) Some metals can be extracted from their ores by reduction with carbon, producing the metal and carbon dioxide.

  (i) Explain the meaning of reduction.

(1 mark)

  (ii) Write a word equation for the reduction of zinc oxide by carbon.

(1 mark)

(c) Iron can be extracted by the reduction of iron(III) oxide (Fe_2O_3) with carbon monoxide (CO), to produce iron and carbon dioxide.

Write a balanced symbol equation for this reaction, including state symbols.

(3 marks)

(d) In which of these test tubes will a reaction occur?

(1 mark)

A 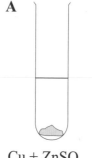   B    C    D

$Cu + ZnSO_{4\,(aq)}$ $Fe + Na_2SO_{4\,(aq)}$ $Zn + CuSO_{4\,(aq)}$ $Fe + ZnSO_{4\,(aq)}$

Exam Questions

2   Which of these statements best describes aluminium?

A   A high density, corrosion resistant metal.

B   One of the main components in steel.

C   A very tough, completely unreactive, dense metal.

D   A light, versatile metal that can't be extracted by reduction with carbon.

(1 mark)

3   Alloys are often used instead of pure metals because

A   they are more plentiful.

B   their properties make them more suitable for the application.

C   their melting points are higher.

D   they are completely inert.

(1 mark)

4   (a)   Use words from the list to label the diagram, which shows the electrolysis process for purifying copper.

anode   cathode   electrical supply   pure copper deposit

(4 marks)

(b)   (i)   Write a balanced half-equation for the reaction at the cathode.

(2 marks)

(ii)   Draw some copper ions on the diagram, with arrows showing their direction of movement. Explain why they move this way.

(3 marks)

(c)   Explain why it is important, economically and environmentally, to recycle copper.

(4 marks)

Revision Summary for Section Four

There wasn't anything too ghastly in this section, and a few bits were even quite interesting, I reckon. But you've got to make sure the facts are all firmly embedded in your brain and that you really understand them. These questions will let you see what you know and what you don't. If you get stuck on any, you need to look at that stuff again. Keep going till you can do them all without coming up for air.

1) Sketch an atom. Label the nucleus and the electrons.

2) Name the three types of particle in an atom. State the relative mass and charge of each particle.

3) What are the symbols for: a) calcium, b) carbon, c) sodium?
   (Use the periodic table at the front of the book.)

4) The element boron is written as $^{11}_{5}B$. How many neutrons do atoms of this element contain? How many electrons does a neutral boron atom have in its outer shell?

5) Describe the difference between a mixture and a compound.

6) Compounds and mixtures are both equally difficult to separate out — true or false?

7)* What atoms make up a molecule of Na_2CO_3?

8)* Say which of the diagrams on the right show:
   a) a mixture of compounds, b) a mixture of elements,
   c) an element, d) a compound

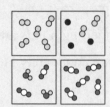

9) Explain how Mendeleev arranged the known elements in a table. How did he predict new elements?

10)*Which element's properties are more similar to magnesium's: calcium or iron?

11)*Balance these equations: a) $CaCO_3 + HCl \rightarrow CaCl_2 + H_2O + CO_2$ b) $Ca + H_2O \rightarrow Ca(OH)_2 + H_2$

12) Write a balanced equation for the reaction between potassium and water.

13) Rubidium ($^{86}_{37}Rb$) is a Group 1 element. When placed in water, a lump of rubidium violently explodes.

   a) Name the gas that is produced by this reaction.

   b) Describe how the pH of the water changes during the experiment.

   c) The experiment is repeated with the same sized lump of sodium ($^{23}_{11}Na$). How would it be different?

14) Describe how the reactivity of Group 7 elements changes as the atomic number increases.

15) Atoms of Group 7 elements tend to go round in pairs. What word describes this type of molecule?

16) Name three different transition metals and give an everyday use for each.

17) Describe how the structure of a metal allows it to carry an electric current.

18) What's the definition of an ore? If something is not an ore now, could it ever be one? Explain...

19) Can magnesium be extracted from its ore by heating with carbon monoxide? If not, why not?

20) What happens if you put:
   a) a piece of magnesium in a solution of zinc sulphate?
   b) a copper bracelet in a solution of iron chloride?

21) What method is used to purify copper once it's been extracted from its ore? Why is this method used?

22) What is the problem with using: a) iron straight from the blast furnace, b) very pure iron?

23) Why are alloys harder than pure metals? Give two examples of alloys and say what's in them.

24) What's so clever about smart alloys?

*Answers on page 291.

Limestone

The Mendip Hills and the Yorkshire Dales are mainly made of a rock called limestone. When limestone is dug out of the ground it's great for building stuff like houses and churches from. You need limestone to make mortar, cement, concrete and glass too. In fact, it's blooming marvellous.

Limestone is used as a **building material**

1) Limestone is a bit of a boring grey/white colour. It's often formed from sea shells and, although the original shells are mostly crushed, there are still quite a few fossilised shells remaining.

2) It's quarried out of the ground. This causes some environmental problems though — see next page.

St Paul's Cathedral is made from limestone.

3) It's great for making into blocks for building with. Fine old buildings like cathedrals are often made purely from limestone blocks. It's also used for statues and fancy carved bits on nice buildings too.

4) Limestone's virtually insoluble in plain water. But acid rain is a big problem. The acid reacts with the limestone and dissolves it away.

5) Limestone can also be crushed up into chippings and used in road surfacing.

Limestone is mainly calcium carbonate

1) Limestone is mainly calcium carbonate — $CaCO_3$. When it's heated it thermally decomposes (breaks down) to make calcium oxide (quicklime) and carbon dioxide.

$$\textbf{calcium carbonate} \rightarrow \textbf{calcium oxide} + \textbf{carbon dioxide}$$

$$CaCO_3(s) \quad \rightarrow \quad CaO(s) \quad + \quad CO_2(g)$$

When other carbonates are heated, they decompose in the same way (e.g. $Na_2CO_3 \rightarrow Na_2O + CO_2$).

2) When you add water to quicklime you get slaked lime. Slaked lime is actually calcium hydroxide.

$$\textbf{quicklime + water} \rightarrow \textbf{slaked lime} \qquad \text{or} \qquad CaO + H_2O \rightarrow Ca(OH)_2$$

3) Slaked lime is an alkali which can be used to neutralise acid soils in fields. Powdered limestone can be used for this too, but the advantage of slaked lime is that it works much faster.

Limestone

Limestone is really very handy. However, digging huge amounts of limestone out of the ground can have a quite a significant negative effect on the environment.

Limestone is used to make other building materials too

1) Powdered limestone is <u>heated</u> in a kiln with <u>powdered clay</u> to make <u>cement</u>.

2) Cement can be mixed with <u>sand</u> and <u>water</u> to make <u>mortar</u>. <u>Mortar</u> is the stuff you stick <u>bricks</u> together with. You can also use it to cover outside walls.

3) Or you can mix cement with <u>sand</u>, <u>water</u> and <u>gravel</u> to make <u>concrete</u>. And by including <u>steel rods</u>, you get <u>reinforced concrete</u> — a <u>composite</u> material with the <u>hardness of concrete</u> and the <u>strength of steel</u>.

4) And believe it or not — limestone is also used to make <u>glass</u>. You just heat it with <u>sand</u> and <u>sodium carbonate</u> until it melts.

Extracting rocks can cause environmental damage

1) Quarrying <u>uses up land</u> and destroys habitats. It costs <u>money</u> to make quarry sites look <u>pretty again</u>. And the <u>waste materials</u> from mines and quarries produce <u>unsightly tips</u>.

2) <u>Transporting rock</u> can cause <u>noise and pollution</u>, and the quarrying process itself produces <u>dust</u> and makes a lot of <u>noise</u> — they often use dynamite to blast the rock out of the ground.

3) Disused sites can be <u>dangerous</u>. <u>Disused mines</u> have been known to <u>collapse</u>. And <u>quarries</u> are sometimes turned into (very deep) lakes — people drown in them every year.

Limestone's amazingly useful

It sounds like you can achieve <u>pretty much anything</u> with limestone. Fred Flintstone even managed to make his car wheels and bowling balls out of rock (although I'm not 100% certain it was limestone).

Useful Products from Air and Salt

The Earth supplies us with pretty much <u>everything</u> we ever use. And until we learn to go into space and mine stuff from asteroids (or whatever), we're going to be <u>absolutely reliant</u> on it in the future as well. Yep... insignificant though it may be on a cosmic scale, I reckon the Earth is pretty amazing.

Fractional distillation of *air* produces *nitrogen* and *oxygen*

1) Air is made up mainly of two gases — <u>nitrogen</u> (about 78%) and <u>oxygen</u> (about 21%). The other 1% is mainly <u>argon</u>. But there's also <u>carbon dioxide</u>, varying amounts of <u>water vapour</u>, and other gases.

2) Nitrogen and oxygen can be <u>separated</u> by <u>fractional distillation</u> — this relies on the fact that the two gases <u>boil</u> at different temperatures (nitrogen at –196 °C, and oxygen at –187 °C).

3) Air is first <u>cooled</u> by repeatedly <u>compressing and expanding</u> it — until it <u>liquefies</u> at about –200 °C. As you 'warm up' the liquefied air, nitrogen boils off first. Easy.

<u>Oxygen</u> is used for oxy-acetylene <u>welding</u>; in <u>rocket engines</u>; to make <u>other chemicals</u>; in <u>hospitals</u>...

<u>Nitrogen</u> is used, e.g. to make <u>ammonia</u> (which is used to make <u>fertilisers</u>), and to provide an <u>unreactive</u> atmosphere. And <u>liquid nitrogen</u> is used to keep things <u>cold</u>; in the drilling of <u>crude oil</u>...

Salt is taken from the *sea* — and from *underneath Cheshire*

1) In <u>hot</u> countries they just pour sea water into big open tanks and let the Sun <u>evaporate</u> the water to leave <u>salt</u>. This is no good in <u>cold</u> countries (like Britain, as if you need reminding) — not enough sunshine.

2) In Britain, salt is extracted from <u>underground deposits</u> left millions of years ago when <u>ancient seas</u> evaporated. There are massive deposits of this rock salt in <u>Cheshire</u>.

3) Rock salt is a mixture of mainly <u>sand</u> and <u>salt</u>. It can be used in its <u>raw state</u> on roads, or the sand can be <u>filtered</u> out of the mixture to leave refined salt.

- The salt in the mixture melts ice by lowering the freezing point of water to around –5 °C.

- The sand and grit give grip on unmelted ice.

4) <u>Refined salt</u> is added to most <u>processed foods</u> to enhance the <u>flavour</u>. (But it's now reckoned to be <u>unhealthy</u> to eat too much salt.)

Useful Products from Air and Salt

Salt is used for making chemicals

1) Salt's also important for the chemical industry.

2) The first thing they do is electrolyse sodium chloride solution (salt water) — this involves passing an electric current through it. It splits the solution into hydrogen, chlorine and sodium hydroxide.

3) Sodium can be extracted by the electrolysis of molten sodium chloride (not sodium chloride solution).

4) All these products can be collected, and then used in all sorts of industries.

Chlorine

1) Used in bleach, for sterilising water, for making hydrochloric acid and insecticides.

2) For making plastics (e.g. it's the C in PVC), pesticides, weedkillers, pharmaceuticals...

Hydrogen

1) Used in the Haber process to make ammonia.

2) Used to change oils into fats for making margarine ('hydrogenated vegetable oil').

3) Used as a fuel in fuel cells, and for welding and metal cutting.

Sodium

1) Sodium's used to make detergents, and lots of other organic chemicals.

2) It's also used in the extraction of titanium, and as a coolant in some nuclear reactors.

Sodium Hydroxide

(Old name: caustic soda)
Sodium hydroxide's a very strong alkali and is used widely in the chemical industry to make, e.g. soap, ceramics, organic chemicals, paper pulp, oven cleaner, 'drain unblocker', bleach... It's also used in aluminium extraction.

Learn the many uses of salt water

Seawater can even be desalinated (have the salt removed) to provide fresh water — but the process needs energy. Eeeh... air, seawater, and rocks under the ground... they give us loads. Don't forget it.

Warm-Up and Exam Questions

By exam time you'll need to know all the facts covered in these questions like the back of your hand. What's more, you'll only get good marks if you can <u>apply</u> the facts — so get practising.

Warm-up Questions

1) Give three major uses of limestone.
2) Give an example of environmental damage caused by quarrying.
3) Describe the difference between cement and mortar.
4) Name two constituents of air other than oxygen and nitrogen.
5) Give two reasons why putting rock salt on icy roads can make driving safer.

Exam Questions

1   Match words **A**, **B**, **C** and **D** with the numbers **1 - 4** in the sentences below.

   A   hydrochloric acid

   B   sodium

   C   sodium hydroxide

   D   hydrogen

   Molten sodium chloride can be electrolysed to give ...**1**...

   The electrolysis of sodium chloride solution produces hydrogen, chlorine and ...**2**...

   Chlorine can be used to make ...**3**..., plastics and weedkillers.

   ...**4**... is used in the Haber process to make ammonia.

   (4 marks)

2   Limestone is mainly calcium carbonate, $CaCO_3$. When heated, it thermally decomposes to produce calcium oxide and carbon dioxide.

   (a)   Write a balanced symbol equation for this reaction.

   (1 mark)

   (b)   Calcium oxide is also known as quicklime. When water is added to quicklime, slaked lime is produced.

      (i)   Write the chemical name and formula of slaked lime.

      (2 marks)

      (ii)   Give **one** use of slaked lime.

      (1 mark)

3   The nitrogen and oxygen in the air can be separated by fractional distillation.

   (a)   What difference between the two gases does this method depend on?

   (1 mark)

   (b)   Give **two** uses of nitrogen.

   (2 marks)

Fractional Distillation of Crude Oil

Crude oil is formed from the buried remains of plants and animals — it's a fossil fuel. Over millions of years, with high temperature and pressure, the remains turn to crude oil, which can be drilled up.

Crude oil can be split into separate hydrocarbons

1) Crude oil is a mixture of hydrocarbons — molecules which are made of just carbon and hydrogen.

2) Fractional distillation splits crude oil into fractions (groups of compounds with carbon chains of similar length).

3) Heated crude oil is piped in at the bottom of a fractionating column.
   The various fractions are constantly tapped off at the different levels where they condense.

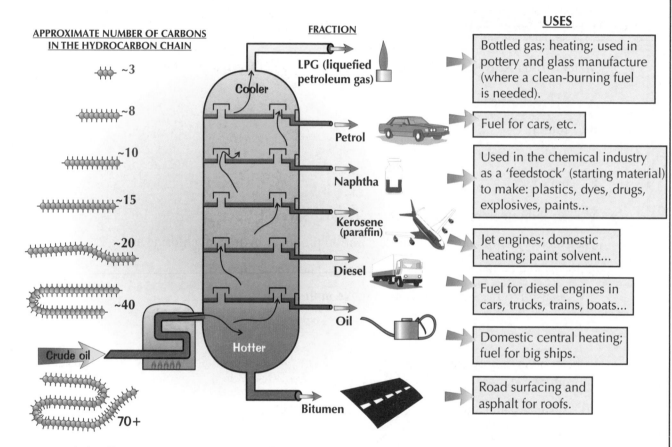

APPROXIMATE NUMBER OF CARBONS IN THE HYDROCARBON CHAIN

~3
~8
~10
~15
~20
~40
70+

Crude oil

Cooler

Hotter

FRACTION

LPG (liquefied petroleum gas)

Petrol

Naphtha

Kerosene (paraffin)

Diesel

Oil

Bitumen

USES

Bottled gas; heating; used in pottery and glass manufacture (where a clean-burning fuel is needed).

Fuel for cars, etc.

Used in the chemical industry as a 'feedstock' (starting material) to make: plastics, dyes, drugs, explosives, paints...

Jet engines; domestic heating; paint solvent...

Fuel for diesel engines in cars, trucks, trains, boats...

Domestic central heating; fuel for big ships.

Road surfacing and asphalt for roofs.

Fractional distillation is an example of a physical process — there are no chemical reactions.

The properties of hydrocarbon molecules depend on their size

The big hydrocarbon molecules are the first to condense, because they have higher boiling points. As the molecules get smaller, they condense higher up the fractionating column. The smaller the molecule...

1) The lower the boiling point — the substance stays as a gas at lower temperatures.

2) The more flammable it is — it sets fire more easily.

3) The less viscous it is — it's less 'gloopy' and flows more easily.

4) The more volatile it is — it evaporates more readily.

The vapours of the more volatile hydrocarbons are very flammable and pose a serious fire risk. So don't smoke at the petrol station. (In fact, don't smoke at all, it's stupid.)

Alkanes

Crude oil contains lots of alkanes and some alkenes (see next page).
They have different properties, and it's all down to their structure.

Alkanes have all C–C single bonds

1) Alkanes are made up of chains of carbon atoms joined by single covalent bonds, and surrounded by hydrogen atoms.

Covalent bonds are when atoms share electrons. Carbon atoms like to make 4 bonds altogether. Hydrogen atoms like to make 1.

2) Different alkanes have chains of different lengths.
The first four alkanes are methane (natural gas), ethane, propane and butane.

---

1) **Methane**: CH_4
(natural gas)

$$H-\overset{\displaystyle H}{\underset{\displaystyle H}{C}}-H$$

---

2) **Ethane**: C_2H_6

$$H-\overset{\displaystyle H}{\underset{\displaystyle H}{C}}-\overset{\displaystyle H}{\underset{\displaystyle H}{C}}-H$$

---

3) **Propane**: C_3H_8

$$H-\overset{\displaystyle H}{\underset{\displaystyle H}{C}}-\overset{\displaystyle H}{\underset{\displaystyle H}{C}}-\overset{\displaystyle H}{\underset{\displaystyle H}{C}}-H$$

---

4) **Butane**: C_4H_{10}

$$H-\overset{\displaystyle H}{\underset{\displaystyle H}{C}}-\overset{\displaystyle H}{\underset{\displaystyle H}{C}}-\overset{\displaystyle H}{\underset{\displaystyle H}{C}}-\overset{\displaystyle H}{\underset{\displaystyle H}{C}}-H$$

---

3) All alkanes have the formula: C_nH_{2n+2}

4) They're called saturated hydrocarbons because they have no spare bonds left (i.e. no double bonds that can open up and have things join onto them — see the next page).

5) You can tell the difference between an alkane and an alkene by adding the substance to bromine water. An alkane won't decolourise the bromine water. This is because it has no spare bonds, so it can't react with the bromine.

 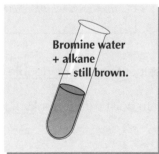

Bromine water
+ alkane
— still brown.

6) Alkanes won't form polymers — again, no spare bonds.

7) They burn cleanly, producing carbon dioxide and water.

Alkenes

Alkenes have very different properties to alkanes. Due to their double bonds they can do lots of clever things — like forming polymers.

Alkenes have a C=C double bond

1) Alkenes have <u>chains</u> of carbon atoms with one or more <u>double covalent bonds</u>.

2) They're called <u>unsaturated hydrocarbons</u> because double bonds can <u>open up</u> and let things join on.

3) This is why they <u>will</u> decolourise <u>bromine water</u>. They form <u>bonds</u> with the bromine.

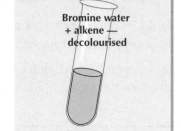

Bromine water + alkene — decolourised

4) <u>Alkenes</u> are <u>more reactive</u> — due to the <u>double bond</u> all poised and ready to just <u>pop open</u>. They can form <u>polymers</u> by <u>opening up</u> their double bonds to '<u>hold hands</u>' in a long chain. (See page 156 for more info on polymers.)

5) The first three alkenes are <u>ethene</u>, <u>propene</u> and <u>butene</u>...

A double bond means that atoms are sharing two pairs of electrons. A double bond counts as two of a carbon atom's four bonds.

---

1) Ethene: C_2H_4

$$\begin{array}{c} H \quad\quad\quad H \\ \backslash/ \\ C=C \\ /\backslash \\ H \quad\quad\quad H \end{array}$$

2) Propene: C_3H_6

$$\begin{array}{c} H \quad H \\ | \quad\; | \\ H-C-C=C \\ | \quad\quad\;\; \backslash \\ H \end{array}$$
H ... H

---

3) Butene: C_4H_8

$$H-C-C=C-C-H \quad \text{or} \quad C=C-C-C-H$$

There are two different forms of butene — the double bond can be in different places.

---

6) <u>All alkenes</u> containing <u>one double bond</u> have the formula: $\quad C_nH_{2n}$

7) They tend to burn with a <u>smoky flame</u>, producing <u>soot</u> (carbon).

Alkane anybody who doesn't learn this lot properly

Don't get alkenes confused with alkanes — that one letter makes all the difference.
Alkenes have a C=C bond, alkanes don't. The first part of their names is the same though.
"Meth-" means "<u>one</u> carbon atom", "eth-" means "<u>two</u> C atoms" , "prop-" means "<u>three</u> C atoms", "but-" means "<u>four</u> C atoms", etc.

Cracking Crude Oil

After the distillation of crude oil, you've still got both short and long hydrocarbons, just not all mixed together. But there's <u>more demand</u> for some products, like <u>petrol</u>, than for others.

Cracking means *splitting up* long-chain hydrocarbons...

1) <u>Long-chain hydrocarbons</u> form <u>thick gloopy liquids</u> like <u>tar</u> which aren't all that useful, so...

2) ... a lot of the longer molecules produced from <u>fractional distillation</u> are <u>turned into smaller ones</u> by a process called <u>cracking</u>.

3) Some of the products of cracking are useful as fuels, e.g. petrol for cars and paraffin for jet fuel.

4) Cracking also produces short alkenes like <u>ethene</u>, which are needed for <u>making plastics</u> (see p156).

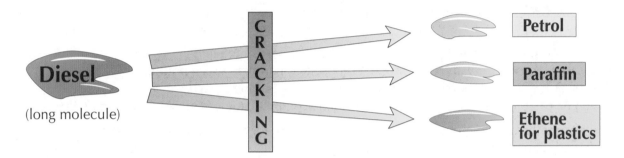

...by passing *vapour* over a hot *catalyst*

1) <u>Cracking</u> is a <u>thermal decomposition</u> reaction — <u>breaking molecules down</u> by <u>heating</u> them.

2) The first step is to <u>heat</u> the long-chain hydrocarbon to <u>vaporise</u> it (turn it into a gas).

3) Then the <u>vapour</u> is passed over a <u>powdered catalyst</u> at a temperature of about <u>400 °C – 700 °C</u>.

4) <u>Aluminium oxide</u> is the catalyst used.

5) The <u>long-chain</u> molecules <u>split apart</u> or "crack" on the <u>surface</u> of the specks of catalyst.

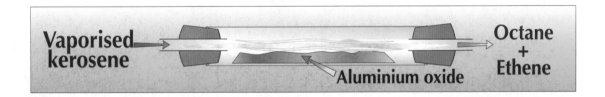

6) Most of the <u>products</u> of cracking are <u>alkanes</u> and <u>alkenes</u> (see pages 125 and 126).

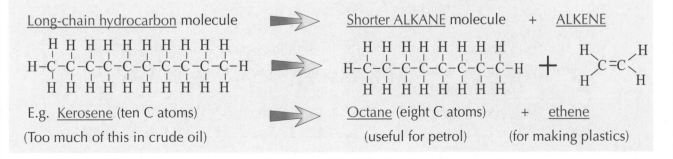

Burning Hydrocarbons

A <u>fuel</u> is a substance that <u>reacts with oxygen</u> to <u>release useful energy</u>. Remember that.

Complete combustion happens when there's plenty of *oxygen*

The <u>complete combustion</u> of any hydrocarbon in oxygen will produce only <u>carbon dioxide</u> and <u>water</u> as waste products, which are both quite <u>clean</u> and <u>non-poisonous</u>.

hydrocarbon + oxygen ➡ carbon dioxide + water   (+ energy)

1) Many <u>gas heaters</u> release these <u>waste gases</u> into the room, which is perfectly OK.
   As long as the gas heater is <u>working properly</u> and the room is <u>well ventilated</u> there's no problem.

2) This reaction, when there's plenty of <u>oxygen</u>, is known as <u>complete combustion</u>.
   It releases <u>lots of energy</u> and only produces those two <u>harmless waste products</u>
   (lots of CO_2 isn't ideal, but the alternatives are worse — see next page).
   When there's <u>plenty of oxygen</u> and combustion is complete, the gas burns with a <u>clean blue flame</u>.

3) You need to be able to give a <u>balanced symbol equation</u> for the complete combustion of a simple hydrocarbon fuel when you're given its <u>molecular formula</u>. It's pretty easy — here's an example:

$$CH_4 + 2O_2 \rightarrow 2H_2O + CO_2 \quad (+ \text{ energy})$$

You've just got to make sure you end up with the <u>same number</u> of Cs, Hs and Os on <u>either side</u> of the arrow.

You can show a fuel burns to give CO_2 and H_2O...

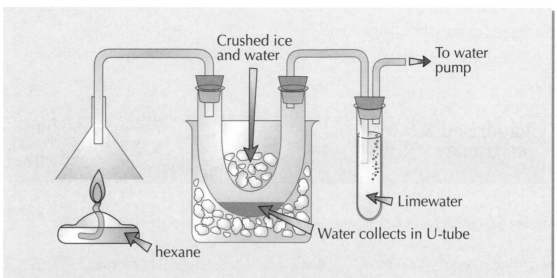

Crushed ice and water

To water pump

Limewater

Water collects in U-tube

hexane

- The <u>water pump</u> draws gases from the burning hexane through the apparatus.
- <u>Water</u> collects inside the <u>cooled U-tube</u> and you can show that it's water by checking its <u>boiling point</u>.
- The <u>limewater turns milky</u>, showing that <u>carbon dioxide</u> is present.

Burning Hydrocarbons

Incomplete combustion also releases useful energy from fuel — but a lot less energy than is released from complete combustion.

Incomplete combustion of hydrocarbons is NOT safe

1) If there isn't enough oxygen the combustion will be incomplete.
   This gives carbon monoxide and carbon as waste products too, and produces a smoky yellow flame.

Incomplete combustion produces less energy than complete combustion does.

hydrocarbon + oxygen ➡

carbon dioxide + water + carbon monoxide + carbon   (+ energy)

2) The carbon monoxide is a colourless, odourless and poisonous gas and it's very dangerous (see p146).
   Every year people are killed while they sleep due to faulty gas fires and boilers filling the room with
   deadly carbon monoxide (CO) and nobody realising — so it's important to regularly service gas
   appliances. The black carbon given off produces sooty marks — a clue the fuel's not burning fully.

3) So basically, you want lots of oxygen when you're burning fuel — you get more energy given out,
   and you don't get any messy soot or poisonous gases.

4) You need to be able to write a balanced symbol equation for incomplete combustion too, e.g.

$$4CH_4 + 6O_2 \rightarrow C + 2CO + CO_2 + 8H_2O$$ (+ energy)

This is just one possibility. The products depend
on the exact quantity of the reactants present...

... E.g. you could also have: $2CH_4 + 3O_2 \rightarrow 2CO + 4H_2O$ —
the important thing is that the equation is balanced.

Blue flame good, yellow flame bad

This is why people should get their gas appliances serviced every year, and get carbon monoxide
detectors fitted. Carbon monoxide really can kill people in their sleep — scary stuff. Don't let that
scare you off learning everything that's on this page — any of it could come up in the exam.

Using Crude Oil as a Fuel

Nothing as amazingly useful as crude oil would be without its problems.
No, that'd be too good to be true.

Crude oil provides an important fuel for modern life

1) Crude oil fractions burn cleanly so they make good <u>fuels</u>.
Most modern transport is fuelled by a crude oil fraction,
e.g. cars, boats, trains and planes. Parts of crude oil are
also burned in <u>central heating systems</u> in homes and in
<u>power stations</u> to <u>generate electricity</u>.

2) There's a <u>massive industry</u> with scientists working
to find oil reserves, take it out of the ground, and
turn it into useful products. As well as fuels,
crude oil also provides the raw materials for
making various <u>chemicals</u>, including <u>plastics</u>.
There's more on this on page 156.

3) Often, <u>alternatives</u> to using crude oil fractions as fuel are
possible. E.g. electricity can be generated by <u>nuclear</u> power
or <u>wind</u> power, <u>solar</u> energy can be used to heat water (see
p209-213), and there are <u>hydrogen</u>-powered cars (see page 148).

4) But things tend to be <u>set up</u> for using oil fractions. For example,
cars are designed for <u>petrol or diesel</u> and it's <u>readily available</u>.
There are filling stations all over the country, with storage facilities
and pumps specifically designed for these crude oil fractions.
So crude oil fractions are often the <u>easiest and cheapest</u> thing to use.

5) Crude oil fractions are often <u>more reliable</u> too — e.g. solar and
wind power won't work without the right weather conditions.
Nuclear energy is reliable, but there are lots of concerns about
its <u>safety</u> and the storage of radioactive waste.

Using Crude Oil as a Fuel

Crude oil is really useful fuel that we use every day — there is a possibility that it might run out.

Crude oil might **run out** one day... eeek

1) Most scientists think that oil will <u>run out</u>. But no one knows exactly when.

2) There have been heaps of <u>different predictions</u> — e.g. about 40 years ago, scientists predicted that it'd all be gone by the year 2000.

3) <u>New oil reserves</u> are discovered from time to time — e.g. a major new oil field was found in southern Oman in the Middle East in 2002. No one knows <u>how much</u> oil will be discovered in the future though.

4) Also, <u>technology</u> is constantly improving, so it's now possible to extract oil that was once too <u>difficult</u> or <u>expensive</u> to extract. It's likely that technology will improve further — but who knows how much?

5) In the <u>worst-case scenario</u>, oil may be pretty much gone in about 25 years — and that's not far off.

6) Some people think we should <u>immediately stop</u> using oil for things like transport, for which there are alternatives, and keep it for things that it's absolutely <u>essential</u> for, like some chemicals and medicines.

7) It will take time to <u>develop</u> alternative fuels that will satisfy all our energy needs (see page 148 for more info). It'll also take time to <u>adapt things</u> so that the fuels can be used on a wide scale. E.g. we might need different kinds of car engines, or special storage tanks built.

8) So however long oil does last for, it's a good idea to start <u>conserving</u> it and finding <u>alternatives</u> now.

Crude oil is **not** the **environment's** best friend

1) <u>Oil spills</u> can happen as the oil is being transported by tanker — this spells <u>disaster</u> for the local environment. <u>Birds</u> get covered in the stuff and are <u>poisoned</u> as they try to clean themselves. Other creatures, like <u>sea otters</u> and <u>whales</u>, are poisoned too.

2) You have to <u>burn oil</u> to release the energy from it. But burning oil is thought to be a major cause of <u>global warming</u> (p 143), <u>acid rain</u> (p 145) and <u>global dimming</u> (p 146).

If oil alternatives aren't developed, we might get caught short

Crude oil is <u>really important</u> to our lives. Take <u>petrol</u> for instance — at the first whisper of a shortage, there's mayhem. Loads of people dash to the petrol station and start filling up their tanks. This causes a queue, which starts everyone else panicking. I don't know what they'll do when it runs out totally.

Warm-Up and Exam Questions

You must be getting used to the routine by now — the warm-up questions run over the basic facts, then the exam questions show you the kind of thing you'll get on the day.

Warm-up Questions

1) What are hydrocarbons?
2) Name three fractions obtained from crude oil.
3) Why are alkenes described as unsaturated hydrocarbons?
4) What sort of hydrocarbon molecules are cracked, and why are they cracked?
5) Describe the conditions used for cracking hydrocarbons.
6) List three modern-day activities that depend on crude oil or its fractions.

Exam Questions

1 The bonds in alkanes are best described as

   A carbon-carbon single bonds and carbon-hydrogen single bonds

   B carbon-carbon single bonds and carbon-hydrogen double bonds

   C carbon-carbon double bonds and carbon-hydrogen single bonds

   D carbon-carbon double bonds and carbon-hydrogen double bonds

   (1 mark)

2 Crude oil can be separated into a number of different compounds as shown in the diagram:

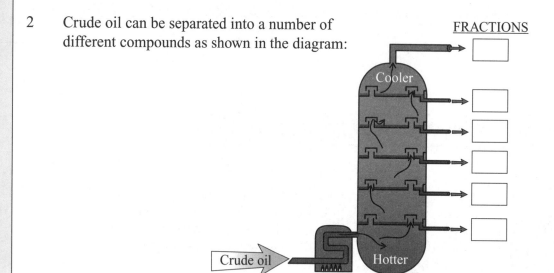

   (a) (i) Put an **M** in the box of the fraction with the largest hydrocarbon molecules.

   (1 mark)

   (ii) Put a **B** in the box of the fraction with the lowest boiling point.

   (1 mark)

   (b) Give **one** use for the kerosene fraction.

   (1 mark)

   (c) Briefly explain how the separation process works.

   (3 marks)

Exam Questions

3   A test for an alkene is

    A   universal indicator is decolourised

    B   starch solution goes dark blue

    C   bromine water is decolourised

    D   bromine water remains brown

(1 mark)

4   Even though there are many environmental problems caused by using crude oil fractions, we continue to use them mainly because

    A   they are a renewable resource

    B   technology is always improving

    C   they are a readily available and concentrated energy source

    D   global warming is only a theory

(1 mark)

5   (a)   When a hydrocarbon fuel burns completely, it produces CO_2 and water.

    (i)   Write a balanced symbol equation for the complete combustion of ethane, C_2H_6.

(2 marks)

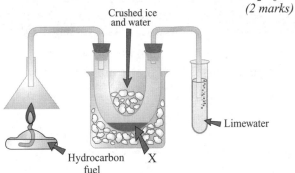

Crushed ice and water

Limewater

Hydrocarbon fuel   X

    (ii)   The apparatus shown can be used to identify the two products of complete combustion of a hydrocarbon, if a further test is carried out on the liquid X which collects in the U-tube.

Describe this test, giving the result you would expect, and explain how the other product is identified.

(2 marks)

    (b)   (i)   What causes incomplete combustion?

(1 mark)

    (ii)   Name **two** products of incomplete combustion that are **not** produced in complete combustion.

(2 marks)

6   A renewable energy source for electricity production is

    A   sunlight

    B   LPG (liquefied petroleum gas)

    C   naphtha

    D   nuclear fuel

(1 mark)

The Earth's Structure

It's tricky to study the structure of the Earth — you can't just dig down to the Earth's centre. But after studying the evidence, this is what scientists think is down there...

Crust, mantle, outer and inner core

1) The crust is Earth's thin outer layer of solid rock. There are two types of crust — continental crust (forming the land), and oceanic crust (under oceans).

2) The lithosphere includes the crust and upper part of the mantle below, and is made up of a jigsaw of 'plates'. The lithosphere is relatively cold and rigid.

3) The mantle extends from the crust almost halfway to the centre of the Earth. It's got all the properties of a solid but it can flow very slowly.

4) The core is just over half the Earth's radius. It's mostly iron and nickel, and is where the Earth's magnetic field originates.

5) The inner core is solid, while the outer core is liquid.

6) Radioactive decay creates a lot of the heat inside the Earth.

7) This heat causes convection currents, which cause the plates of the lithosphere to move (which is bad news for some people — see below).

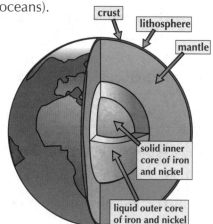

The Earth's surface is made up of tectonic plates

1) The crust and the upper part of the mantle are cracked into a number of large pieces called tectonic plates. These plates are a bit like big rafts that 'float' on the mantle.

2) The plates don't stay in one place though. That's because the convection currents in the mantle cause the plates to drift. The map shows the edges of the plates as they are now, and the directions they're moving in (red arrows).

3) Most of the plates are moving at speeds of a few cm per year relative to each other.

4) Occasionally, the plates move very suddenly, causing an earthquake. Volcanoes often form at the boundaries between two tectonic plates too.

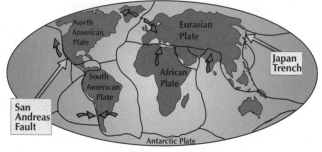

Scientists can't predict earthquakes and volcanic eruptions

1) Tectonic plates can stay more or less put for a while and then suddenly lurch forwards. It's impossible to predict exactly when they'll move.

2) Scientists are trying to find out if there are any clues that an earthquake might happen soon — things like strain in underground rocks. Even with these clues they'll only be able to say an earthquake's likely to happen, not exactly when it'll happen.

3) There are also clues that a volcanic eruption might happen soon. Before an eruption, molten rock rises up into chambers near the surface, causing the ground surface to bulge slightly. This causes mini-earthquakes near the volcano.

4) But sometimes molten rock cools down instead of erupting, so mini-earthquakes can be a false alarm.

Evidence for Plate Tectonics

A bloke called Alfred Wegener put forward his theory about the Earth's continents slowly drifting along in 1915, but not many people believed it. This was partly because he didn't have a good explanation for <u>why</u> it happened, partly because he wasn't a qualified <u>geologist</u>, and partly because the theory was so <u>weird</u>. But the truth will out, as they say — and the evidence now suggests the 'rocky raft' idea is correct.

1) *Jigsaw fit* — *the supercontinent 'Pangaea'*

a) There's a very obvious <u>jigsaw fit</u> between <u>Africa</u> and <u>South America</u>.

b) The <u>other continents</u> can <u>also</u> be fitted in without too much trouble.

c) It's widely believed that they once all formed <u>a single land mass</u>, now called <u>Pangaea</u>.

2) *Matching fossils* in Africa and South America

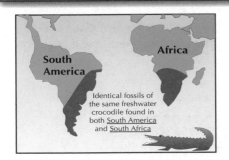

a) Identical <u>plant fossils</u> of the <u>same age</u> have been found in rocks in <u>South Africa</u>, <u>Australia</u>, <u>Antarctica</u>, <u>India</u> and <u>South America</u>, which strongly suggests they were all <u>joined</u> once upon a time.

b) <u>Animal fossils</u> support the theory too. There are identical fossils of a <u>freshwater crocodile</u> found in both <u>Brazil</u> and <u>South Africa</u>. It certainly didn't swim across.

3) *Identical* rock sequences

a) Certain <u>rock layers</u> of similar <u>ages</u> in various countries show remarkable <u>similarity</u>.

b) This is strong evidence that these countries were <u>joined together</u> when the rocks <u>formed</u>.

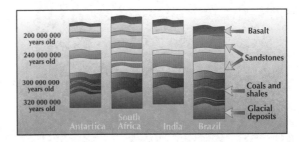

4) *Living creatures*: The Earthworm

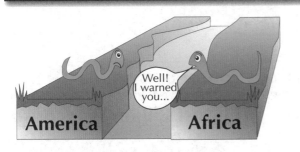

a) There are various <u>living creatures</u> found in <u>both</u> America and Africa.

b) One such beasty is a particular <u>earthworm</u> which is found living at the <u>tip of South America</u> and the <u>tip of South Africa</u>.

c) Most likely it travelled across <u>ever so slowly</u> on the big raft we now call America.

Learn about Wegener's Theory and all the evidence

So there you go. Alfred Wegener's ideas were originally thought to be <u>bonkers</u> (the fact that he'd used some <u>inaccurate data</u> didn't help — one scientist claimed that the forces Wegener's theory needed would have stopped the Earth rotating). But as <u>technology improved</u> and more evidence was gathered (including from the bottom of the ocean), it turned out that Wegener's ideas were pretty convincing after all. But it took a while — it was only in the 1960s that scientists really accepted the theory.

Evolution of the Atmosphere

For 200 million years or so, the atmosphere has been about how it is now: 78% nitrogen, 21% oxygen, and small amounts of other gases, mainly CO_2 and noble gases. There can be a lot of water vapour too. But it wasn't always like this. Here's how the past 4.5 billion years may have gone:

Phase 1 — *Volcanoes gave out gases*

1) The Earth's surface was originally <u>molten</u> for many millions of years. It was so hot that any atmosphere just '<u>boiled away</u>' into space.

2) Eventually things cooled down a bit and a <u>thin crust</u> formed, but <u>volcanoes</u> kept erupting.

3) The volcanoes gave out lots of gas — including <u>carbon dioxide</u>, <u>water vapour</u> and <u>nitrogen</u>. We think this was how the oceans and atmosphere were formed.

4) According to this theory, the early atmosphere was probably <u>mostly CO_2</u>, with virtually <u>no oxygen</u>. This is quite like the atmospheres of Mars and Venus today.

5) The <u>oceans</u> formed when the water vapour <u>condensed</u>.

The First Two Billion Years

Steam CO_2 N_2 N_2

<u>Holiday report</u>: Not a nice place to be. Take strong walking boots and a good coat.

Phase 2 — *Green plants evolved and produced oxygen*

The Next Two Billion Years

O_2 O_2 O_2 O_2 O_2 O_2 O_2 O_2 O_2 O_2

<u>Holiday report</u>: A bit slimy underfoot. Take wellies and a lot of suncream.

1) <u>Green plants</u> evolved over most of the Earth. They were quite happy in the <u>CO_2 atmosphere</u>.

2) A lot of the early CO_2 <u>dissolved</u> into the oceans. The <u>green plants</u> also <u>removed CO_2</u> from the air and <u>produced O_2</u> by <u>photosynthesis</u>.

3) When plants died and were buried under layers of sediment, the <u>carbon</u> they had removed from the air (as CO_2) became 'locked up' in <u>sedimentary rocks</u> as <u>insoluble carbonates</u> and <u>fossil fuels</u>.

4) When we <u>burn</u> fossil fuels today, this 'locked-up' carbon is released and the concentration of CO_2 in the atmosphere rises.

Phase 3 — *Ozone layer allows evolution of complex animals*

1) The build-up of <u>oxygen</u> in the atmosphere <u>killed off</u> some early organisms that couldn't tolerate it, but allowed other, more complex organisms to evolve and flourish.

2) The oxygen also created the <u>ozone layer</u> (O_3) which <u>blocked</u> harmful rays from the Sun and <u>enabled</u> even <u>more complex</u> organisms to evolve — us, eventually.

3) There is virtually <u>no CO_2</u> left now.

<u>Holiday report</u>: A nice place to be. Visit before the crowds ruin it.

The Last Billion Years or so

Nice safe OZONE, O_3

Changes in the Atmosphere

Evidence for how the atmosphere evolved has been found in rocks and other sources. But no one was there to record the changes as they happened. So our ideas about the atmosphere are still theories...

There are **competing theories** about **atmospheric change**

As well as the theory on the previous page, there are <u>other theories</u> about how the Earth's atmosphere changed millions of years ago. Ultimately, all the theories have to be judged on the evidence.

> For example, one theory says that the water on Earth came mainly from <u>comets</u> rather than volcanoes. When this theory was first suggested, it seemed far-fetched. But space science research soon suggested that lots of <u>small icy comets</u> really are hitting the Earth <u>every day</u>. So far so good. But studies of comets found that the water in comets <u>isn't the same</u> as the water on Earth (it's got more 'heavy water' in it). So current thinking is that most of Earth's water <u>probably didn't</u> come from comets.

The atmosphere **changes** all the time

1) This is a graph of <u>CO_2</u> and <u>global temperature</u> data. It shows CO_2 levels <u>rising rapidly</u> over the last few thousand years, and a global temperature rise that's been more or less keeping up.

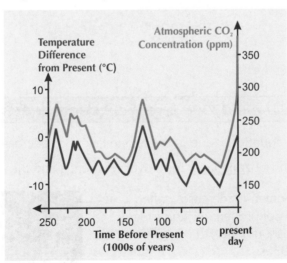

2) <u>But</u> the graph also shows that there have been <u>huge changes</u> in the climate before. (These changes are small beer compared with the changes described on page 136, when the <u>entire composition</u> of the atmosphere was changing... but they're still pretty big.)

3) For instance, there have been several <u>ice ages</u> over the last few million years. These happen for various reasons (e.g. things to do with the Earth's orbit, movement of continents, CO_2 in the atmosphere, and so on).

   An ice age is a time when large areas of the Earth's surface are covered with ice.

4) So <u>changes</u> in the Earth's <u>temperature</u> aren't <u>new</u> — they happen all the time. However, we've recently become aware of the possibility of a <u>faster warming</u> of the planet, and what this could mean for us.

---

4 million years ago was a whole other world

We've learned a lot about the past atmosphere from <u>Antarctic ice cores</u>. Each year, a layer of ice forms and <u>bubbles of air</u> get trapped inside it, then it's buried by the next layer. So the deeper the ice, the older the air — and if you examine the bubbles in different layers, you can see how the air has changed.

Changes in the Atmosphere

There have been huge changes in the climate in the past. Now the climate is changing again, but there's still lots of debate about whether it is just a natural change or change brought about by human activity.

The atmosphere is **still changing**

Levels of CO_2 in the atmosphere are **increasing**...

Levels of CO_2 in the atmosphere have increased by about 25% since 1750...

1) <u>Burning fossil fuels</u> releases CO_2 — and as the world has become more industrialised, more fossil fuels have been burnt in power stations and in car engines.

2) Carbon dioxide is a <u>greenhouse gas</u> — it traps heat from the Sun. You'd expect that more carbon dioxide would mean a <u>hotter</u> planet. (See page 143 for more info.)

3) However, a <u>few</u> scientists say that the concentration of CO_2 has <u>increased and decreased a lot</u> over the last <u>100s of millions of years</u>, and argue that a little increase now might be just a <u>blip</u>.

The amount of **ozone** in the ozone layer has **decreased**...

Over the last 50 years, the amount of ozone in the ozone layer has decreased...

1) Currently, <u>holes</u> in the Earth's <u>ozone layer</u> form over Antarctica and the Arctic each year.

2) Ozone is broken down by man-made gases called <u>CFCs</u>, widely used as aerosol propellants and fridge coolants between the 1930s and the 1980s. CFCs were phased out in the 1990s.

3) The ozone layer protects us from the harmful UV radiation which can cause skin cancer. It's <u>difficult to test</u> whether changes in the ozone layer are to blame for increases in skin cancer, though. <u>Other factors</u> affect skin cancer — people <u>sunbathe</u> more and have more <u>beach holidays abroad</u>, so they expose themselves to more UV radiation anyway.

The atmosphere is changing — it could be all our fault

Whether people believe scientific theories or not depends on the <u>evidence</u> that people produce to support them. Without <u>evidence</u>, a theory goes nowhere. Quite right too, I say.

Warm-Up and Exam Questions

Doing these warm-up questions will soon tell you if you've got the basic facts straight.
If not, you'll really struggle, so take time to go back over the bits you don't know.

Warm-up Questions

1) State one geological feature often seen at the boundary of two tectonic plates.
2) What was 'Pangaea'?
3) How do scientists believe the oceans formed?
4) What process releases heat inside the Earth?
5) Give an example of a visible clue that a volcano might be about to erupt.
6) Give one effect of the build-up of oxygen in the atmosphere over the last billion years.

Exam Questions

1   Match words **A**, **B**, **C** and **D** with the numbers **1 - 4** in the sentences below.

   A   photosynthesis

   B   oxygen

   C   carbon

   D   carbon dioxide

   Once green plants had evolved, they thrived in an atmosphere rich in ...**1**...
   These plants produced ...**2**... by the process of ...**3**...
   ...**4**... from dead plants eventually became 'locked up' in fossil fuels.

   (4 marks)

2   The following graph shows how
     atmospheric CO_2 concentration and
     global temperature have varied over
     the last 250 000 years.

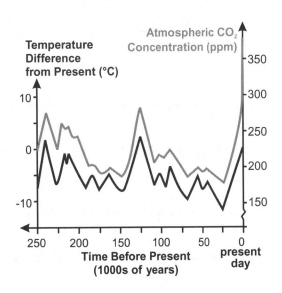

   (a)   Describe what the graph shows about temperature and CO_2 levels.

   (2 marks)

   (b)   Mark with an X on the graph the time when the temperature was most different
          from its present value.

   (1 mark)

Exam Questions

3   Tectonic plates generally move at speeds of

    A   several metres per year

    B   between 5 and 10 kilometres per hour

    C   a few centimetres per year

    D   1 or 2 millimetres per century

(1 mark)

4   The existence of animal fossils of the same species in both South Africa and Brazil suggests that

    A   these continents are moving towards each other

    B   South America and Africa were once joined together

    C   South America and Africa were never joined together

    D   South America and Africa have changed places

(1 mark)

5   The following diagram shows the internal structure of the Earth.

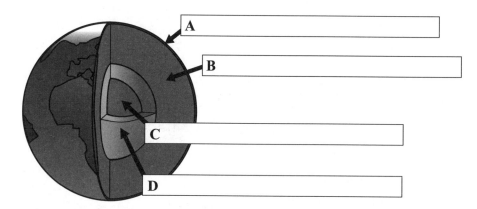

    (a)   Label the diagram.

(4 marks)

    (b)   What **two** elements do scientists believe part **C** is largely made from?
        Put a ring around the correct answers.

        cadmium   nickel   silicon   aluminium   iron

(2 marks)

6   (a)   (i)   Name two regions over which the Earth's ozone layer is thinner than normal.

(2 marks)

        (ii)   What effect do CFCs (chlorofluorocarbons) have on ozone?

(1 mark)

        (iii)   Give **one** use of CFCs.

(1 mark)

    (b)   Discuss the possible connection between ozone levels and a rise in the incidence of skin cancer in humans.

(3 marks)

Human Impact on the Environment

We have an impact on the world around us — and the more humans there are, the bigger the impact.

There are **six billion people** in the world...

1) The population of the world is currently rising very quickly, and it's not slowing down
   — look at the graph...

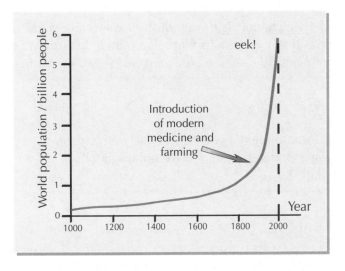

2) This is mostly due to modern medicine and farming methods, which have reduced the number of
   people dying from disease and hunger.

3) This is great for all of us humans, but it means we're having a bigger effect on the environment
   we live in...

...with **increasing demands** on the **environment**

When the Earth's population was much smaller, the effects of human activity were usually small
and local. Nowadays though, our actions can have a far more widespread effect.

1) Our rapidly increasing population puts pressure on the
   environment, as we take the resources we need to survive.

2) But people around the world are also demanding a higher
   standard of living (and so demand luxuries to make life more
   comfortable — cars, computers, etc.). So we use more raw
   materials (e.g. oil to make plastics), and we also use more
   energy for the manufacturing processes. This all means we're
   taking more and more resources from the environment more
   and more quickly.

3) Unfortunately, many raw materials are being used up quicker
   than they're being replaced. So if we carry on like we are,
   one day we're going to run out.

Human Impact on the Environment

We are damaging the environment in several different ways...

We're also producing **more waste**

As we make more and more things we produce more and more <u>waste</u>. And unless this waste is properly handled, more <u>harmful pollution</u> will be caused. This affects water, land and air.

Water
<u>Sewage</u> and <u>toxic chemicals</u> from industry can pollute lakes, rivers and oceans, affecting the plants and animals that rely on them for survival (including humans). And the chemicals used on land (e.g. fertilisers) can be washed into water.

Land
We use <u>toxic chemicals</u> for farming (e.g. pesticides and herbicides). We also bury <u>nuclear waste</u> underground, and we dump a lot of <u>household waste</u> in landfill sites.

Air
<u>Smoke</u> and <u>gases</u> released into the atmosphere can pollute the air (see pages 143, 145 and 146 for more). For example, <u>sulphur dioxide</u> can cause <u>acid rain</u>.

More people means **less land** for plants and other animals

Humans also <u>reduce</u> the amount of <u>land and resources</u> available to other <u>animals</u> and <u>plants</u>. The <u>four main human activities</u> that do this are:

1) <u>Building</u>

2) <u>Dumping Waste</u>

4) <u>Quarrying</u>

3) <u>Farming</u>

More people, more mess, less space, less resources

In the exam you might be given some data about <u>environmental impact</u>, so make sure you understand what's going on. Just keep your head and work out exactly what the data's saying. Job's a good 'un.

Global Warming

Most environmentalists and scientists now believe that <u>human activities</u> are <u>changing</u> the proportion of <u>carbon dioxide</u> in the <u>atmosphere</u> — and that that's going to have <u>massive effects</u> on life on Planet Earth. You need to understand the <u>science</u> behind the scary headlines — starting with the <u>greenhouse effect</u>.

Carbon dioxide and methane trap heat from the Sun

1) The <u>temperature</u> of the Earth is a <u>balance</u> between the heat it gets from the Sun and the heat it radiates back out into space.

2) Gases in the <u>atmosphere</u> absorb most of the heat that would normally be radiated into space, and re-radiate it in all directions (including towards Earth). If this didn't happen, then at night there'd be nothing to keep any heat <u>in</u>, and we'd quickly get <u>very cold</u> indeed. But recently we've started to worry that this effect is getting a bit out of hand...

3) There are several different gases in the atmosphere which help keep the <u>heat in</u>. They're called "<u>greenhouse gases</u>" (oddly enough) and the <u>main ones</u> whose levels we worry about are <u>carbon dioxide</u> and <u>methane</u> — because the levels of these two gases are rising quite sharply.

4) <u>Humans</u> release <u>carbon dioxide</u> into the atmosphere as part of our <u>everyday lives</u> — e.g. as we <u>burn fossil fuels</u> in power stations or cars.

5) This could be a big problem, but it's hard to be 100% sure (since Earth's climate is so complicated). For example, the Earth's temperature <u>varies</u> over the years anyway (see page 137) — so even if Earth is warming up, it <u>might</u> be nothing to do with humans and fossil fuels.

But nowadays, <u>most</u> scientists think that:

(i) Earth <u>is</u> gradually warming

(ii) fossil fuel use <u>has</u> got something to do with it.

The Carbon Cycle

Carbon flows through the Earth's ecosystem in the carbon cycle.
This cycle can help us see why the amount of carbon dioxide in the atmosphere has increased.

Carbon is constantly being recycled

Carbon is the key to the greenhouse effect — it exists in the atmosphere as <u>carbon dioxide gas</u>, and is also present in many other <u>greenhouse gases</u> such as methane.

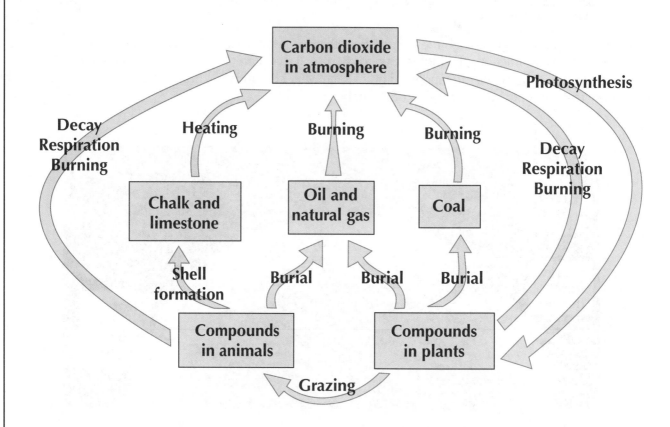

1) The carbon on Earth moves in a big cycle.

2) <u>Respiration</u> (see page 24), <u>combustion</u> (see page 128) and <u>decay</u> of plants and animals add carbon dioxide to the air and remove oxygen.

3) <u>Photosynthesis</u> (see page 78) does the <u>opposite</u>
   — it removes carbon dioxide and adds oxygen.

4) These processes should <u>balance out</u>.
   However, it looks like <u>humans</u> have upset the natural carbon cycle.

Eeeek — the carbon cycle's got a puncture

Releasing masses of CO_2 due to <u>fossil fuel use</u> is the worst problem here. But <u>deforestation</u> (cutting down large areas of forest) is also bad for the carbon cycle, in <u>three</u> ways. First off, <u>carbon dioxide</u> is <u>released</u> when trees are <u>burnt</u> to clear land. Secondly, <u>microorganisms</u> feeding on the dead wood <u>release CO_2</u> due to respiration. And thirdly, fewer trees means there's <u>less photosynthesis</u> going on.

Air Pollution

Carbon dioxide levels could be causing climate change. But CO_2 isn't the only gas released when fossil fuels burn — you also get other nasties like oxides of nitrogen and sulphur dioxide.

Acid rain is caused by sulphur dioxide and oxides of nitrogen

1) When fossil fuels are burned they release mostly CO_2.
But they also release other harmful gases — e.g. sulphur dioxide and various nitrogen oxides.

2) The sulphur dioxide (SO_2) comes from sulphur impurities in the fossil fuels.

3) However, the nitrogen oxides are created from a reaction between the nitrogen and oxygen in the air, caused by the heat of the burning. (This can happen in the internal combustion engines of cars.)

4) When these gases mix with clouds they form dilute sulphuric acid and dilute nitric acid. This then falls as acid rain.

5) Power stations and internal combustion engines in cars are the main causes of acid rain.

Acid rain kills fish, trees and statues

1) Acid rain causes lakes to become acidic and many plants and animals die as a result.

2) Acid rain kills trees and damages limestone buildings and ruins stone statues. It also makes metal corrode. It's shocking.

Oxides of nitrogen also cause photochemical smog

1) Photochemical smog is a type of air pollution caused by sunlight acting on oxides of nitrogen.

2) These oxides combine with oxygen in the air to produce ozone (O_3).

3) Ozone can cause breathing difficulties, headaches and tiredness.
(Don't confuse ground-level ozone with the useful ozone layer high up in the atmosphere.)

Acid rain — bad for the trees

But good for you, because acid rain can get you some nice easy marks in the exam. Make sure you remember the names of the two gases which cause acid rain, and where they come from.

Air Pollution

Carbon monoxide is another pollutant that can be released from burning fossil fuels — it's pretty nasty. Pollution in the atmosphere such as CO can cause health problems, so it needs to be carefully controlled.

Carbon monoxide is a poisonous gas

1) Carbon monoxide (CO) can stop your blood doing its proper job of carrying oxygen around the body.

2) A lack of oxygen in the blood can lead to fainting, a coma or even death.

3) Carbon monoxide is formed when petrol or diesel in car engines is burnt without enough oxygen — this is incomplete combustion (see page 129 for more details).

It's important that atmospheric pollution is controlled

1) The build-up of all these pollutants can make life unhealthy and miserable for many humans, animals and plants. The number of cases of respiratory illnesses (e.g. asthma) has increased in recent years — especially among young people. Many people blame atmospheric pollution for this, so efforts are being made to improve things.

2) Catalytic converters on motor vehicles reduce the amount of carbon monoxide and nitrogen oxides getting into the atmosphere. The catalyst is normally a mixture of platinum and rhodium. It helps unpleasant exhaust gases from the car react to make things that are less immediately dangerous (though more CO_2 is still not exactly ideal).

3) And Flue Gas Desulphurisation (FGD) technology in some fossil-fuel power stations removes sulphur dioxide from the exhaust gases.

$$\text{carbon monoxide} + \text{nitrogen oxide} \rightarrow \text{nitrogen} + \text{carbon dioxide}$$
$$2CO + 2NO \rightarrow N_2 + 2CO_2$$

Global dimming — caused by burning fossil fuels (maybe)

1) Over the last few years, some scientists have measured how much sunlight is reaching Earth.

2) They've found that in some areas nearly 25% less sunlight has been reaching the surface compared to 50 years ago. They have called this global dimming.

3) They think it's caused by particles of soot and ash that are produced when fossil fuels are burnt. These particles reflect sunlight back into space, or they can help to produce more clouds that reflect the sunlight back into space. (There are many scientists who don't believe that the change is real though, and blame it on inaccurate recording equipment.)

London by night. (And by day too if global dimming gets really bad.)

Protecting the Atmosphere

What to do... what to do... How can we look after the atmosphere...

Computer models are used to make predictions

1) <u>Computer models</u> are used to predict the temperature of the Earth's atmosphere in the future. They use data collected by <u>thousands</u> of <u>monitoring stations</u> all over the world.

2) The data is fed into the models, then millions of <u>calculations</u> are carried out.

3) However, computer models are only as good as the <u>data</u> you put into them, and the <u>assumptions</u> made when working out the calculations. <u>If</u> the assumptions are <u>wrong</u>, this could lead to <u>false results</u>. And <u>one small error</u> in an early calculation could be <u>magnified</u> if it's used to predict further into the future. (Having said that, early comparisons of <u>computer predictions</u> and <u>observed events</u> look pretty good.)

The precautionary principle — better safe than sorry

1) Various governments have agreed to apply the '<u>precautionary principle</u>' to climate change.

2) The idea is that we should <u>assume the worst</u> and therefore <u>reduce</u> CO_2 emissions. If we turn out to be <u>wrong</u>, then the climate's <u>safe</u> anyway. But if we turn out to be <u>right</u>, we've taken <u>early action</u>.

3) There are two basic strategies for combating climate change...

(i) Burn less fossil fuels.
   We should burn fossil fuels <u>more efficiently</u>, and also use <u>other sources</u> of energy (that don't emit greenhouse gases — e.g. nuclear, wind, etc.).

(ii) Burn fossil fuels but try to stop levels of greenhouse gases increasing so much. We could <u>capture</u> some of the CO_2 <u>before</u> it's released into the atmosphere. And we could <u>plant forests</u> to <u>absorb</u> some of the CO_2 (as they photosynthesise — see page 78).

Revision and pollution — the two bugbears of modern life
Eeee.... <u>cars</u> and <u>fossil fuels</u> — they're nowt but trouble. But at least this topic is kind of interesting, what with its relevance to everyday life and all. Just think... you could see this kind of stuff on TV.

Protecting the Atmosphere

As the demand on resources increases it is important to develop new, alternative fuels.

Alternative fuels are being developed

Some alternatives to fossil fuels already exist, and there are others in the pipeline (so to speak).
They should reduce the amount of fossil fuels burnt.

> **BIOGAS** is a mixture of methane and carbon dioxide. It's produced when
> microorganisms digest waste material. It can be produced on a large scale,
> or on a small scale where each family has its own generator.
> Biogas is burned and the energy can be used for cooking, heating or lighting.
>
> PROS: Waste material is readily available and cheap. It's 'carbon neutral'.
>
> CONS: Biogas production is slow in cool weather.

And see page 175 about using ethanol as a fuel.

> **HYDROGEN GAS** can also be used to power vehicles. You get the hydrogen from
> the electrolysis of water. There's plenty of water about but it takes electrical energy
> to split it up — however, this energy can come from a renewable source, e.g. solar.
>
> PROS: Hydrogen combines with oxygen in the air to form just water — so it's very clean.
>
> CONS: You need a special, expensive engine and hydrogen isn't widely available.
> You still need to use energy from another source to make it.
> Also, hydrogen's hard to store — it's very explosive.

There's lots to consider when choosing a fuel

1) Energy value (i.e. amount of energy) — funnily enough, this isn't always as important as it may seem.

2) Availability — there's not much point in choosing a fuel you can't get hold of easily.

3) Storage — some fuels take up a lot of space, and some produce flammable gases.

4) Cost — some fuels are expensive, but still good value in terms of energy content etc.

5) Toxicity — poisonous fumes are a problem.

6) Pollution — e.g. will you be adding to acid rain and the greenhouse effect? Or causing lots of smoke?

Example: You're at home and there's a power cut. You want a cup of tea. The only fuels you have in
the house are candles or meths (in a spirit burner). Which one would you use to boil the water?

Fuel	Energy per gram	Rate of energy produced	Flame
Meths	28 kJ	15 kJ per minute	Clean
Candle	50 kJ	8 kJ per minute	Smoky

Even though a candle has more energy
per gram, you'd probably choose meths
because it's quicker and cleaner.

I produce a kind of biogas already

As it's mostly fossil fuel use that gives the atmosphere such a hard time, alternative fuels are a good
thing to start looking for. And getting people to use less energy is also a pretty sensible idea too.

Warm-Up and Exam Questions

You could just read through this page thinking 'yep, I can do that', but there'd not be much point. You need to write down an answer to every question, check them, and look up anything you didn't know.

Warm-up Questions

1) Give one reason why the world's population is rising rapidly.
2) How is the rise in our standard of living affecting the environment?
3) Carbon dioxide is a greenhouse gas. What does this mean?
4) Describe two effects of acid rain.
5) Name two gases, apart from CO_2, that are often produced when fossil fuels burn.
6) Why is hydrogen considered to be a 'clean fuel'?
7) Give four important criteria to consider when choosing a fuel.

Exam Questions

1   The large human population is currently leading to

   A   a reduction in quarrying

   B   less pollution

   C   rapid depletion of some resources

   D   more land being available for plants

   (1 mark)

2   The diagram below shows the carbon cycle.

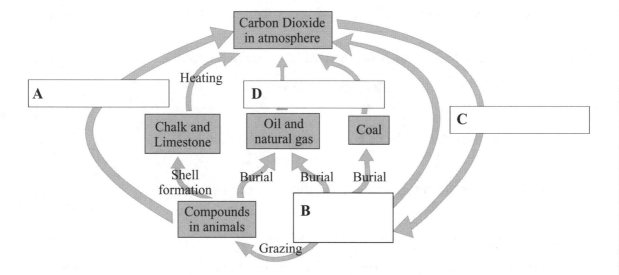

   (a)   Choose from the words and phrases in the list to fill in the blanks in the diagram.

   respiration   compounds in plants   burning   photosynthesis   burial

   (4 marks)

   (b)   Describe how large-scale deforestation affects CO_2 concentration in the atmosphere, and explain why.

   (3 marks)

Exam Questions

3   Global warming happens when

    A   more of the Sun's heat is reflected into space than is absorbed by the Earth

    B   oceans absorb more and more CO_2

    C   there is more CO_2 and methane in the atmosphere

    D   we use a greater proportion of renewable energy sources

(1 mark)

4   Match words **A**, **B**, **C** and **D** with the numbers **1 - 4** in the sentences below.

    A   carbon monoxide

    B   nitrogen oxides

    C   oxygen

    D   smog

    Acid rain is formed when ...**1**... and SO_2 mix with clouds.

    Nitrogen oxides can cause photochemical ...**2**....

    Breathing in another pollutant, ...**3**..., hinders the uptake of ...**4**... by the blood and is potentially fatal.

(4 marks)

5   Currently, fossil fuels provide about 60-70% of the world's electricity.

    Since fossils fuels will eventually run out, it is important to find alternative energy sources for the future.

    (a) Biogas is one fuel that may be more widely used in the future.

        (i)   What are the main constituents of biogas?

(1 mark)

        (ii) How is biogas produced and what could it be used for?

(2 marks)

        (iii) Give **one** advantage and **one** disadvantage of using biogas.

(2 marks)

    (b) Another 'fuel for the future' could be hydrogen.

        (i)   Describe how hydrogen is produced on a large scale.

(1 mark)

        (ii) Why is hydrogen difficult to store?

(1 mark)

        (iii) Give **one** other disadvantage of using hydrogen as a fuel in a car compared with using petrol or diesel.

(1 mark)

Biodiversity and Indicator Species

Human activity is having a negative effect on the environment — this could have an impact on the animals that share the planet. Destroying habitat and polluting the environment could cause the extinction of species. Some animals can help us monitor our impact on the environment.

Reduction in biodiversity could be a big problem

Biodiversity is the variety of different species present in an area — the more species, the higher the biodiversity. Ecosystems (especially tropical rainforests) can contain a huge number of different species, so when a habitat is destroyed there is a danger of many species becoming extinct — biodiversity is reduced. This causes a number of lost opportunities for humans and problems for those species that are left:

1) There are probably loads of useful products that we will never know about because the organisms that produced them have become extinct. Newly discovered plants and animals are a great source of new foods, new fibres for clothing and new medicines, e.g. the rosy periwinkle flower from Madagascar has helped treat Hodgkin's disease (a type of cancer), and a chemical in the saliva of a leech has been used to help prevent blood clots during surgery.

2) Loss of one or more species from an ecosystem unbalances it, e.g. the extinct animal's predators may die out or be reduced. Loss of biodiversity can have a 'snowball effect' which prevents the ecosystem providing things we need, such as rich soil, clean water, and the oxygen we breathe.

Human impact can be measured using indicator species

Getting an accurate picture of the human impact on the environment is hard. But one technique that's used involves indicator species.

1) Some organisms are very sensitive to changes in their environment and so can be studied to see the effect of human activities — these organisms are known as indicator species.

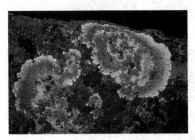

2) For example, air pollution can be monitored by looking at particular types of lichen, which are very sensitive to levels of sulphur dioxide in the atmosphere (and so can give a good idea about the level of pollution from car exhausts, power stations, etc.). The number and type of lichen at a particular location will indicate how clean the air is (e.g. the air is clean if there are lots of lichen).

3) And if raw sewage is released into a river, the bacterial population in the water increases and uses up the oxygen. Animals like mayfly larvae are good indicators for water pollution, because they are very sensitive to the level of oxygen in the water. If you find mayfly larvae in a river, it indicates that the water is clean.

Teenagers are an indicator species — not found in clean rooms

In the exam, make sure you remember the details about the environmental problems that development can cause. If you get an essay-type question, stick 'em in to show off your 'scientific knowledge'.

Recycling Materials

It's not all doom and gloom... if we do things sustainably e.g. by recycling, we'll be okay.

It's *important* to *recycle*

There are various reasons why...

1) Use less resources

There's a <u>finite amount</u> of materials (e.g. metals, oil for plastics) in the Earth. Recycling <u>conserves</u> these resources.

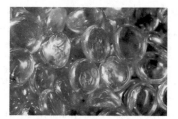

2) Use less energy

Mining, extracting and making materials (e.g. metals or glass) need lots of <u>energy</u>, which mostly comes from burning <u>fossil fuels</u>. Fossil fuels will <u>run out</u> one day, and they also cause <u>pollution</u> (see page 146 for more about <u>global dimming</u>, but also have a look at pages 137-146 for more general pollution info). Recycling things like <u>copper</u>, <u>aluminium</u> and <u>glass</u> takes a <u>fraction</u> of the energy.

3) Use less money

Energy doesn't come cheap, so recycling <u>saves money</u> too.

4) Make less rubbish

Recycling also cuts down on the amount of rubbish that goes to <u>landfill</u>, which takes up space and <u>pollutes</u> the surroundings.

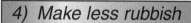

This is a reason why we should recycle paper.

*Hard work never killed anyone, but why take a chance**

You can calculate the <u>financial</u> benefits of recycling any material, but remember there are the '<u>resources</u>', '<u>energy</u>' and '<u>rubbish</u>' benefits too. With <u>paper</u>, <u>sustainable forests</u> are good (where for every tree you cut down, you plant another one), but that doesn't reduce the amount of <u>landfill</u>.

* This is presented here in aid of Recycle an Elderly Joke Week

Recycling Materials

Recycling will benefit humans now and help to reduce our impact on future generations.

*There may be **economic** and **environmental** benefits of recycling*

1) Working out the <u>cost benefits</u> of recycling can get a bit <u>tricky</u> — there's lots to take into account.

2) For example, recycling <u>isn't</u> free. There are <u>costs</u> involved in <u>collecting</u> waste material, <u>transporting</u> it, <u>sorting</u> it, and then <u>processing</u> it.

3) But if you didn't recycle, say, <u>aluminium</u>, you'd have to <u>mine</u> more aluminium ore — <u>4 tonnes</u> for every <u>1 tonne</u> of aluminium you need. But mining makes a mess of the <u>landscape</u> (and these mines are often in <u>rainforests</u>). The ore then needs to be <u>transported</u>, and the aluminium <u>extracted</u> (which uses <u>loads</u> of electricity). And don't forget the cost of sending your <u>used</u> aluminium to <u>landfill</u>.

4) So it's a <u>complex</u> calculation, but for every 1 kg of aluminium cans you recycle, you <u>save</u>:

> - <u>95%</u> or so of the <u>energy</u> needed to mine and extract 'fresh' aluminium,
> - <u>4 kg</u> of aluminium ore,
> - a <u>lot</u> of waste.

In fact, aluminium's about the most cost-effective metal to recycle.

5) But even if all these differences were very <u>small</u>, maybe it's still worth recycling — you're getting people <u>involved</u> in doing their bit for the environment. Can't be a bad thing.

Sustainable development** needs careful **planning

1) Human activities can <u>damage</u> the environment (e.g. pollution). And some of the damage we do can't easily be <u>repaired</u> (e.g. the destruction of the rainforests).

2) We're also placing <u>greater pressure</u> on our planet's <u>limited resources</u> (e.g. oil is a non-renewable resource so it will eventually run out).

3) This means that we need to <u>plan carefully</u> to make sure that our activities today don't mess things up for <u>future generations</u> — this is the idea behind <u>sustainable development</u>...

> <u>SUSTAINABLE DEVELOPMENT</u> meets the needs of <u>today's</u> population <u>without</u> harming the ability of <u>future</u> generations to meet their own needs.

4) This isn't easy — it needs detailed thought at every level to make it happen. For example, <u>governments</u> around the world will need to make careful plans. But so will the people in charge at a <u>regional</u> level.

Recycling is an example of sustainable development

Recycling materials is a great way in which we can start to reduce our impact on future generations. But there are lots of other ways too — e.g. managing forests sustainably and introducing fishing quotas.

Warm-Up and Exam Questions

The warm-up questions run quickly over the basic facts you'll need in the exam. The exam questions come later — but unless you've learnt the facts you'll find the exams tougher than leather sandwiches.

Warm-up Questions

1) Name two non-renewable resources.
2) In a cost-benefit analysis of recycling, name an important cost.
3) Give two benefits of recycling.

Exam Questions

1   (a)   Biodiversity means

     A   the use of genetically modified organisms

     B   allowing certain species to become extinct

     C   artificial selection of characteristics

     D   the number and variety of different species in a location

(1 mark)

    (b)   Explain why a reduction in biodiversity in an ecosystem could be a problem for:

     (i)   remaining species within that ecosystem

(1 mark)

     (ii)   humans

(1 mark)

2   What is meant by 'sustainable development'?

   A   The needs of today's population are met without harming the ability of future generations to meet their needs.

   B   By recyling, all the needs of future generations can be met.

   C   Meeting the energy needs of the population.

   D   The energy needs of today's population are met without harming the ability of future generations to meet their energy needs.

(1 mark)

3   Some organisms can be used to help us judge the level of pollution.

    (a)   What name is given to an organism that can be used in this way?

(1 mark)

    (b)   Different species can be studied to judge the levels of different pollutants in the air or in water.

     (i)   Give **one** example of an organism used to show pollution levels.

(1 mark)

     (ii)   Explain what the organism you chose in part i) can tell us about pollution levels, and how.

(2 marks)

Revision Summary for Section Five

The structure and atmosphere of the Earth, limestone, alkanes, alternative fuels — can they really belong in the same section, I almost hear you ask. Whether you find the topics easy or hard, interesting or dull, you need to learn it all before the exam. Try these questions... see how much you know:

1) How is glass made? Cement? Concrete?

2) Name four substances obtained from seawater or rock salt, and give two uses for each.

3) Explain briefly the principle of fractional distillation.

4) What does crude oil consist of? Draw the full diagram of the fractional distillation of crude oil.

5) Describe four properties of hydrocarbons and how they vary with the molecule size.

6) What's the general formula for an alkane? What's the formula for a 5-carbon alkane?

7) What kind of carbon-carbon bond do alkenes have?
   Give the general formula for an alkene containing one double bond.

8) Draw the chemical structure of ethene.

9) Describe a test you can do to tell whether a particular hydrocarbon is an alkane or an alkene.

10) Give a typical example of a substance that is cracked, and the products that you get from cracking it.

11) Give the equations for complete and incomplete combustion of hydrocarbons.

12) Explain how incomplete combustion can be harmful to humans.

13)*Write down a balanced symbol equation for the incomplete combustion of ethane (C_2H_6).

14) Why do predictions about when crude oil will run out change over the years?

15) Describe two ways in which oil slicks affect wildlife.

16) What is the lithosphere?

17) How does tectonic plate movement cause: a) earthquakes? b) volcanoes?
    What causes the Earth's tectonic plates to move?

18) Describe four pieces of evidence for the theory of plate tectonics.

19) For a long time, the Earth's early atmosphere was mostly CO_2. Where did this CO_2 come from?

20) Name the two main gases that make up the Earth's atmosphere today. What processes produced O_2?

21) Explain how the ozone layer has enabled complex organisms to evolve.

22) Describe two ways in which the atmosphere is changing today.

23) Suggest three ways in which a rising population affects the environment.

24) Name two "greenhouse gases". How do they affect the temperature of the Earth?

25) Sketch and label a diagram of the carbon cycle.

26) Which gases cause "acid rain"? How do they get into the air?
    Describe two ways of reducing acid rain.

27) Describe what is meant by "photochemical smog".

28) Name a poisonous gas that catalytic converters help to remove from car exhausts.

29) Explain what is meant by the precautionary principle, and how it can be applied to global warming.

30) Explain the benefits and difficulties of using hydrogen to power vehicles. Do the same for biogas.

31) What are indicator species? Give examples.

32) Explain how lichen can be used as an indicator of air pollution.

*Answers on page 292.

Polymers

Plastics are made up of lots of molecules joined together. They're like long chains.

Plastics are long-chain molecules called polymers

1) Plastics are formed when lots of small molecules called monomers join together to give a polymer.

2) They're usually carbon based (and the monomers are very often alkenes — see page 126).

Addition polymers are made under high pressure

1) The monomers that make up addition polymers have a double covalent bond (i.e. they're unsaturated).

2) Under high pressure and with a catalyst (see page 189) to help them along, many unsaturated small molecules open up those double bonds and "join hands" (polymerise) to form long saturated chains called polymers.

Ethene becoming polyethene or "polythene", is the easiest example:

$$n\left(\overset{|}{\underset{|}{C}}=\overset{|}{\underset{|}{C}}\right) \longrightarrow \left(\overset{|}{\underset{|}{C}}-\overset{|}{\underset{|}{C}}\right)_n$$

Many single ethenes

Polyethene

The 'n' just means there can be any number of monomers.

You'll need to be able to draw the formula of an addition polymer, given the formula of its monomer. Dead easy — the carbons just all join together in a row with no double bonds between them.

The name of the plastic comes from the type of monomer it's made from — you just stick the word "poly" in front of it:

Propene can form polypropene:

Propene → Polypropene

A molecule called styrene will polymerise into polystyrene:

Styrene → Polystyrene

$\bigcirc = C_6H_5$

Forces between molecules determine the properties of plastics

Strong covalent bonds hold the atoms together in long chains. But it's the bonds between the different molecule chains that determine the properties of the plastic.

Weak Forces:

Long chains held together by weak forces are free to slide over each other. This means the plastic can be stretched easily, and will have a low melting point.

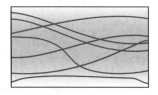

Strong Forces:

Plastics with stronger bonds between the polymer chains have higher melting points and can't be stretched, as the crosslinks hold the chains firmly together.

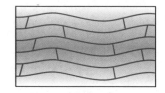

Uses of Polymers

Plastics are fantastically useful. You can make novelty football pencil sharpeners and all sorts.

Polymers' **properties** decide what they're **used for**

Different polymers have different physical properties — some are stronger, some are stretchier, some are more easily moulded, and so on. These different physical properties make them suited for different uses.

- Strong, rigid polymers such as high density polyethene are used to make plastic milk bottles.

- Light, stretchable polymers such as low density polyethene are used for plastic bags and squeezy bottles. Low density polyethene has a low melting point, so it's no good for anything that'll get very hot.

- PVC is strong and durable, and it can be made either rigid or stretchy. The rigid kind is used to make window frames and piping. The stretchy kind is used to make synthetic leather.

- Polystyrene foam is used in packaging to protect breakable things, and it's used to make disposable coffee cups (the trapped air in the foam makes it a brilliant thermal insulator).

- Heat-resistant polymers such as melamine resin and polypropene are used to make plastic kettles.

Non-biodegradable plastics cause disposal problems

1) Most polymers aren't "biodegradable" — they're not broken down by microorganisms, so they don't rot. This property is actually kind of useful until it's time to get rid of your plastic.

2) It's difficult to get rid of plastics — if you bury them in a landfill site, they'll still be there years later. Landfill sites fill up quickly, and they're a waste of land. And a waste of plastic.

3) When plastics are burnt, some of them release gases such as acidic sulphur dioxide and poisonous hydrogen chloride and hydrogen cyanide. So burning's out, really. Plus it's a waste of plastic.

4) The best thing is to reuse plastics as many times as possible and then recycle them if you can. Sorting out lots of different plastics for recycling is difficult and expensive, though.

5) Chemists are working on a variety of ideas to produce biodegradable polymers.

Uses of Polymers

Polymers are often used to make clothes

1) Nylon is a <u>synthetic polymer</u> often used to make clothes. Fabrics made from nylon are not waterproof on their own, but can be coated with <u>polyurethane</u> to make tough, hard-wearing and waterproof outdoor clothing.

2) One big problem is that the polyurethane coating doesn't let <u>water vapour</u> pass through it. So if you get a bit hot (or do a bit of exercise), sweat <u>condenses</u> on the inside. This makes skin and clothes get <u>wet and uncomfortable</u> — the material isn't <u>breathable</u>.

3) <u>Breathable</u> fabrics have all the <u>useful properties of nylon/polyurethane</u> ones, but they also let sweat <u>out</u>. If you sweat in a breathable material, water vapour can <u>escape</u> — so <u>no condensation</u>.

Polymers are used to make breathable materials

1) Some breathable fabrics are made by combining a thin film of a plastic with a layer of another fabric, such as nylon.

2) The plastic film has <u>tiny holes</u> which let <u>water vapour</u> pass through — so it's <u>breathable</u>. But it's <u>waterproof</u>, since the holes aren't big enough to let <u>big water droplets</u> through and the plastic <u>repels liquid water</u>.

3) This material is great for <u>outdoorsy types</u> — they can hike without getting rained on or soaked in sweat.

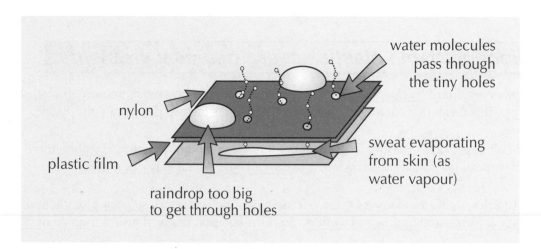

Polymers have a wide range of uses

If you're making a <u>product</u>, you need to pick your plastic carefully. It's no good trying to make a jacket out of a brittle, unbendy polymer — imagine trying to walk any distance in a jacket like that. The same goes for things like kettles — there's no point using a plastic that melts at 50 °C.

Warm-Up and Exam Questions

Warm-up Questions

1) Explain what monomers and polymers are.
2) How can a plastic's melting point tell you about the bonds between its polymer chains?
3) In a polymer molecule, what type of bonds hold the atoms together in a chain?
4) Melamine is a heat-resistant polymer. Suggest a use for melamine.

Exam Questions

1 Complete the passage using some of the following words.

explosive	**biodegradable**	**hidden**	**recycle**
expensive	**toxic**	**flammable**	**burnt**

The majority of plastics are hard to dispose of because they are not

.... They can be ...., but this often

releases .... gases like hydrogen cyanide. Burying plastics

also has its disadvantages. This is why it is best to ....

plastics whenever possible, however this can be an ....

process.

(5 marks)

2 (a) Complete this equation for the formation of polypropene.

$$n \left(\begin{array}{c} H \\ | \\ C \\ | \\ H \end{array} = \begin{array}{c} H \\ | \\ C \\ | \\ CH_3 \end{array} \right) \longrightarrow$$

(1 mark)

(b) The structural formula of polystyrene is shown below.
Draw the structural formula of its monomer.

$$\left(\begin{array}{cc} H & H \\ | & | \\ C & C \\ | & | \\ H & \bigcirc \end{array} \right)_n$$

(1 mark)

(c) PVC (polyvinyl chloride) is strong and durable and can be made in rigid or stretchy forms. Suggest a use for:

(i) the rigid form.

(1 mark)

(ii) the stretchy form.

(1 mark)

Paints and Pigments

Different minerals in the Earth are different colours, and these minerals have been used as pigments for thousands of years. Nowadays of course, you get those fancy paint mixing machines in DIY warehouses.

Pigments give paints their colours

1) Paint usually contains the following bits: solvent, binding medium and pigment.

2) The pigment gives the paint its colour.

3) The binding medium is a liquid that carries the pigment bits and holds them together. When the binding medium turns to solid, it sticks the pigments to the surface you've painted.

4) The solvent is the stuff that keeps the binding medium and pigment runny — as a liquid when it's in the tin or as a paste when it is still in the tube.

Paints are colloids

1) A colloid consists of really tiny particles of one kind of stuff dispersed in (mixed in with) another kind of stuff. They're mixed in, but not dissolved.

2) The particles can be bits of solid, droplets of liquid or bubbles of gas.

3) Colloids don't separate out because the particles are so small. They don't settle out at the bottom.

4) In an oil paint, the pigment is in really tiny bits dispersed in the oil. And then the solvent (if there is one — there isn't always) dissolves the oil to keep it all runny.

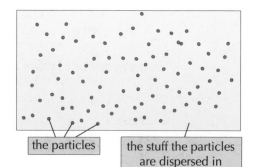

the particles

the stuff the particles are dispersed in

Some paints are water-based and some are oil-based

1) Emulsion paints are water-based. The solvent in the paint is water, and the binding medium is usually a polymer such as polyurethane, acrylic or latex.

2) Traditional gloss paint and artists' oil paints are oil-based. This time, the binding material is oil, and the solvent in the paint is an organic compound that'll dissolve oil. Turpentine is used as a solvent for artists' oil paints. Some solvents in oil-based paints produce fumes which can be harmful — it's best to make sure there's plenty of ventilation when using oil-based gloss.

3) Whether you're creating a masterpiece in oils or painting your bedroom wall, you normally brush on the paint as a thin layer. The paint dries as the solvent evaporates. (A thin layer dries a heck of a lot quicker than a thick layer.)

4) With a water-based emulsion, the solvent evaporates, leaving behind the binder and pigment as a thin solid film. A thin layer of emulsion paint dries quite quickly.

   Some modern gloss paints are water-based.

5) Oil-based paints take rather longer to dry because the oil has to be oxidised by oxygen in the air before it turns solid.

In the exam, you might be asked to choose the best kind of paint for a job, given some info about paints. For example, to paint the outside part of a door you'd want a waterproof, hard-wearing paint. Oil-based paints are more hard-wearing than water-based paints, so you'd probably go for an oil-based gloss. To paint bedroom walls you'd want a paint that goes on easily, dries quickly, and doesn't produce harmful fumes. Water-based emulsion fits the bill here.

Perfumes

Some things smell nice, some don't... it's all down to the <u>chemicals</u> a substance contains.

Perfumes can be natural or artificial

1) Chemicals that smell nice are used as <u>perfumes</u> and <u>air fresheners</u>. <u>Esters</u> are often used as perfumes as they usually smell quite <u>pleasant</u>.

2) Esters are pretty common in <u>nature</u>. Loads of common <u>food smells</u> (plus those in products like <u>perfumes</u>) contain <u>natural esters</u>.

3) Esters are also <u>manufactured</u> <u>synthetically</u> to enhance <u>food flavours</u> or <u>aromas</u>, e.g. there are esters (or combinations of esters) that smell of rum, apple, orange, pineapple, and so on. And esters are responsible for the distinctive smell of <u>pear drops</u>.

Esters are made by esterification

1) <u>Esters</u> can be made by heating a <u>carboxylic acid</u> with an <u>alcohol</u>. (This is an example of <u>esterification</u>.)

A carboxylic acid is an acid built around one or more <u>carbon atoms</u>.

2) An <u>acid catalyst</u> is usually used (e.g. <u>concentrated sulphuric acid</u>).

$$\text{Acid} + \text{Alcohol} \rightarrow \text{Ester} + \text{Water}$$

Learn this equation.

Method:

1) Mix 10 cm³ of a carboxylic acid such as <u>ethanoic acid</u> with 10 cm³ of an alcohol such as <u>ethanol</u>.

2) Add 1 cm³ of <u>concentrated sulphuric acid</u> to this mixture and <u>warm gently</u> for about 5 minutes.

3) Tip the mixture into <u>150 cm³ of sodium carbonate solution</u> (to neutralise the acids) and <u>smell carefully</u> (by wafting the smell towards your nose). The <u>fruity-smelling product</u> is the ester.

Perfumes

Not just any ester will do though...

Perfumes need **certain properties**

You can't use any old chemical with a <u>smell</u> as a perfume. You need a substance with <u>certain properties</u>:

1) <u>Easily evaporates</u> — or else the perfume particles <u>won't</u> reach your nose and you won't be able to smell it... bit useless really.

See page 186 for more about how we smell perfumes.

2) <u>Non-toxic</u> — it mustn't seep through your skin and <u>poison</u> you.

3) <u>Doesn't react with water</u> — or else it would react with the water in <u>sweat</u>.

4) <u>Doesn't irritate the skin</u> — or else you couldn't <u>apply it directly</u> to your neck or wrists. If you splash on any old substance you risk <u>burning</u> your skin.

5) <u>Insoluble in water</u> — if it was soluble in water it would <u>wash off</u> every time you got wet.

Don't forget that even if a substance has <u>all</u> these properties, it still might <u>smell</u> pretty bad and so be <u>unsuitable</u> for a perfume.

New perfumes and **cosmetics** have to be **tested**

Companies are always developing new cosmetic products to sell to us. Before they're released to the shops, they need to be <u>tested thoroughly</u> to make sure they're <u>safe to use</u>. They should be <u>non-toxic</u> and shouldn't <u>irritate</u> the eyes or skin. Pretty obvious, I'm sure you'll agree. But some tests are carried out using animals, which is a bit more controversial.

<u>Advantages</u> of testing new cosmetics on animals: We get an idea of whether they're likely to irritate the skin or be toxic <u>before</u> humans use them (though an animal test won't <u>necessarily</u> apply to humans).

<u>Disadvantages</u> of testing on animals: The tests could cause <u>pain</u> and suffering to the animals (especially if it turns out that the cosmetic <u>is</u> toxic). And animals <u>can't choose</u> whether or not to take part in the tests (so using human volunteers instead could be a possibility in certain circumstances).

Testing on animals is a controversial issue

Perfume needs to <u>smell</u> nice, but not everyone <u>agrees</u> on what smells nice. Perfume also needs to be <u>safe</u>, but not everyone agrees on the <u>best</u> way to test for this. That's life for you.

Fancy Materials

New materials are continually being developed. Their properties determine what they can be used for.

New materials are sometimes invented by accident

1) Teflon® is what the non-stick coating on non-stick saucepans is made from. It was discovered in 1938 — by accident. Scientists were trying to make a new refrigerant and came up with Teflon® instead.

2) The glue on 'Post-it® Notes' was invented when scientists were trying to develop a new glue for sticky tape, and ended up with a glue that didn't stick permanently (so it was no good for sticky tape). Later on, someone came up with the idea of sticky-backed peel-off notes.

3) Some materials are deliberately invented for one particular use and other uses come later. KEVLAR® is a synthetic polymer that's strong, light, flexible, heat-resistant and electrically insulating. It was invented to strengthen car tyres and now it's got loads more uses (e.g. bulletproof vests).

Nanomaterials are really really really really tiny
...smaller than that.

Really tiny particles, 1–100 nanometres (nm) across, are called 'nanoparticles' (1 nm = 0.000 000 001 m). For comparison, a grain of sand is about 0.5 mm across — at least 5000 times bigger than a nanoparticle.

A nanoparticle has very different properties from the 'bulk' chemical it's made from.

1) Nanoparticles of titanium dioxide and zinc oxide are used in some sunscreens. Nanoparticles of titanium dioxide and zinc oxide are so small that they reflect UV light but not visible light. They're actually so small they're invisible. They don't leave any marks.

2) Nanoparticles of silver can stop viruses from getting into cells. Ordinary silver doesn't fight viruses.

Fancy Materials

Smart materials can change their properties

Here are some examples of smart materials that <u>behave differently</u> depending on the <u>conditions</u>.

1) <u>Nitinol</u> is a "<u>shape memory alloy</u>". No matter how much you twist it, it'll bend back into shape when you heat it. See page 115 for more information.

2) Some <u>dyes</u> can change <u>colour</u> or become transparent depending on <u>the temperature</u>. They're used on some <u>drinks cans and bottles</u> to show when the contents are <u>cold enough</u> to drink, and on <u>novelty mugs</u> with designs that change when a hot drink is poured in.

3) And similar materials can be used in food packaging — special dyes which <u>change colour</u> faster the <u>warmer</u> they get can be used to tell if a food has been warm for long enough for <u>microbes</u> to grow.

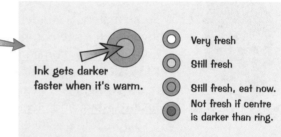

Ink gets darker faster when it's warm.

- Very fresh
- Still fresh
- Still fresh, eat now.
- Not fresh if centre is darker than ring.

4) There are also dyes that become more or less <u>transparent</u> depending on <u>light intensity</u>. They're used in <u>sunglasses</u> that get darker in more intense sunlight.

5) Some materials <u>expand</u> or <u>contract</u> when you put an <u>electric current</u> through them. They also do the <u>opposite</u> — they produce <u>electricity</u> when they're <u>squeezed</u>. These materials are used in <u>car airbag sensors</u>. When squeezed by the forces of a car crash they produce electricity to activate the airbag.

6) There are <u>liquids</u> that turn <u>solid</u> when you put them in a <u>magnetic field</u>. They're used to control vibrations, e.g. in some car <u>shock absorbers</u>.

7) And phosphorescent pigments <u>absorb natural or artificial light</u> and <u>store the energy</u> in their molecules. This energy is <u>released</u> as <u>light</u> over a period of time — from a few seconds to a couple of hours. Obvious uses are signs, toys, novelty decorations, glow-in-the-dark hands on watches, etc.

Learn the properties and uses of these smart materials

For a brief spell in the early 1990s colour-changing T-shirts were reeeeally cool. Although I was never quite sure why you'd want to show the world just how hot and sweaty your armpits were.

Warm-Up and Exam Questions

It's time to practise some more questions. If you struggle with the warm-up questions, do some more revision then try them again. Only do the exam questions when you think you know your stuff properly.

Warm-up Questions

1) Which three components are found in most paints?
2) Why do new perfumes and cosmetics have to be tested?
3) What are phosphorescent pigments? Give one possible use for them.
4) What is the difference between water-based and oil-based paints?
5) Give one possible use of a dye that becomes less transparent with increasing light intensity.

Exam Questions

1   Many paints are colloids.

(a)   What is a colloid?

(1 mark)

(b)   Why don't colloids separate out?

(1 mark)

(c)   Read the following sentences about different paints and answer the questions that follow.

> Modern emulsion paints use water as a solvent. These paints produce very low levels of harmful fumes and are fast drying.
>
> Most gloss paints use organic compounds as solvents. Gloss paint should only be used in well ventilated areas because it can produce harmful fumes. It is harder-wearing than emulsion but takes longer to dry.

(i)   Which type of paint would you use to paint the front door of a house — emulsion or gloss? Explain your answer

(1 mark)

(ii)   Which type of paint would you use to paint the walls of a bedroom? Explain your answer

(1 mark)

2   Esters are often used as perfumes. It is possible to make an ester by reacting a carboxylic acid with an alcohol.

(a)   This is an example of what type of reaction?

(1 mark)

(b)   Write a word equation for the reaction between a carboxylic acid and an alcohol.

(1 mark)

(c)   List five properties that perfumes need to have.

(5 marks)

Chemicals and Food

Cooking is just chemistry by another name — chemistry involving pies.

Some **foods** have to be **cooked**

There are loads of different ways to cook food —

e.g.   boiling,
       steaming,
       grilling,
       frying...

1) Many foods have a better taste and texture when cooked.

2) Some foods are easier to digest once they're cooked (e.g. potatoes, flour). See below for why.

3) The high temperatures involved in cooking also kill off those nasty little microbes that cause disease (see page 47) — this is very important with meat.

4) Some foods are poisonous when raw, and must be cooked to make them edible — e.g. red kidney beans contain a poison that's only destroyed by at least 10 minutes boiling (and 2 hours cooking in total).

Cooking causes **chemical changes**

Cooking food produces new substances. That means a chemical change has taken place. Once cooked, you can't change it back. The cooking process is irreversible.

e.g. Eggs and meat

Eggs and meat are good sources of protein. Protein molecules change shape when you heat them. The energy from cooking breaks some of the chemical bonds in the protein (see page 185), and this allows the molecule to take a different shape. This gives the food a more edible texture. The change is called denaturing — it's irreversible.

e.g. Potatoes

Potatoes are a good source of carbohydrates. Potatoes are plants, so each potato cell is surrounded by a cellulose cell wall. Humans can't digest cellulose, so this makes it difficult to get to the contents of the cells. Cooking the potato breaks down the cell wall, making it a lot easier to digest.

Chemicals and Food

Emulsifiers Help Oil and Water Mix

1) You can mix an oil with water to make an <u>emulsion</u>. Emulsions are made up of lots of <u>droplets</u> of one liquid <u>suspended</u> in another liquid.

2) Oil and water <u>naturally separate</u> into two layers with the oil floating on top of the water — they don't "want" to mix. <u>Emulsifiers</u> help to stop the two liquids in an emulsion from <u>separating out</u>.

3) <u>Mayonnaise</u>, <u>low-fat spread</u> and <u>ice cream</u> are foods which contain emulsifiers.

4) Emulsifiers are molecules with one part that's <u>attracted to water</u> and another part that's <u>attracted to oil</u> or fat. The bit that's attracted to water is called <u>hydrophilic</u>, and the bit that's attracted to oil is called <u>hydrophobic</u>.

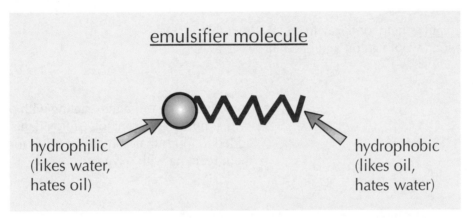

emulsifier molecule

hydrophilic
(likes water,
hates oil)

hydrophobic
(likes oil,
hates water)

5) The <u>hydrophilic</u> end of each emulsifier molecule latches onto <u>water molecules</u>.

6) The <u>hydrophobic</u> end of each emulsifier molecule cosies up to <u>oil molecules</u>.

7) When you shake oil and water together with a bit of emulsifier, the oil forms droplets, surrounded by a coating of emulsifier... <u>with the hydrophilic bit facing outwards</u>. Other oil droplets are <u>repelled</u> by the hydrophilic bit of the emulsifier, while water molecules latch on. So the emulsion won't separate out. Clever.

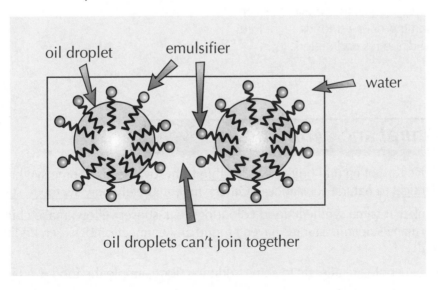

oil droplet   emulsifier

water

oil droplets can't join together

Cooking food makes it easier to digest and safer

When you cook something, you're bringing about chemical change. The changes are irreversible, as you'll know if you've ever tried to <u>unscramble</u> an egg.

Food Additives

Humans have been adding stuff to food for years. Before fridges were invented, we added salt to meat to stop it going off. Now we use additives not just to preserve food, but to make it look or taste different.

Processed foods often contain additives

1) Food manufacturers add various chemical compounds to food to improve its appearance, taste, texture and shelf life. These additives must be listed in the ingredients list on the back of the packet.

2) Most additives used in the UK have E-numbers — e.g. E127 is erythrosine (a red dye) and E201 is sodium sorbate (a preservative). Additives with E-numbers have passed safety tests and can be used in Europe.

- Preservatives help food stay fresh. Without them, more food would go off and need throwing away.

- Some foods 'go off' after reacting with oxygen — e.g. butter goes rancid. Antioxidants are added to foods that contain fat or oil to stop them reacting with oxygen.

- Colourings and flavourings make food look and taste better.

- Emulsifiers (see previous page) and stabilisers stop emulsions like mayonnaise from separating out.

- Sweeteners can replace sugar in some processed foods — helpful to diabetics and dieters.

There are natural and synthetic additives

1) Some food additives are of natural origin, e.g. lecithin from soya beans. Some synthetic additives are identical to natural substances. Others are completely new synthetic substances.

2) Some people think that some synthetic food colourings (e.g. sunset yellow) make children hyperactive. But many scientific studies haven't found any connection between additives and hyperactivity at all.

3) A small number of people are allergic to some additives, for example the food dye tartrazine.

4) Some additives aren't suitable for vegetarians. For example, the food colouring cochineal comes from crushed insects. And gelatin from animal bones is used to thicken and set some foods.

Food Additives

Artificial colours can be detected by chromatography

To identify different colourings in a food sample, you can use chromatography.

> Paper chromatography uses the fact that different dyes wash through wet filter paper at different rates.

Here's how you'd analyse food colourings...

1) Extract the colour from each food sample by placing it in a small cup with a few drops of solvent (can be water, ethanol, salt water etc). Use a different cup for each different food sample.

2) Put spots of each coloured solution on a pencil baseline on filter paper. (Label them in pencil — don't use pen because it might dissolve in the solvent and confuse everything.)

3) Roll up the sheet and put it in a beaker with some solvent — but keep the baseline above the level of the solvent.

4) The solvent seeps up the paper, taking the food dyes with it. Different dyes form spots in different places.

5) Watch out though — a chromatogram with four spots means at least four dyes, not exactly four dyes. There could be five dyes, with two of them making a spot in the same place. (It can't be three dyes though, because one dye can't split into two spots.)

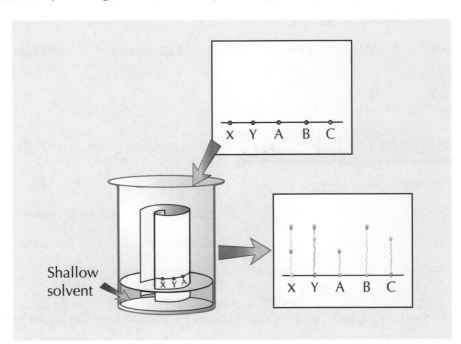

Learn all the different additives and why they're added to food

Chromatography can separate even complex mixtures if you choose the right equipment and conditions. On a different note, there's a lot about food additives in the media. Some statements are based on facts, others on rumour and prejudice. But without evidence to support a claim, it's not worth a bean.

Plant Oils in Food

Plant oils come from plants. I know it's tricky, but just do your best to remember.

We can **extract oils** from **plants**

1) Some fruits and seeds contain a lot of oil. For example, avocados and olives are oily fruits. Brazil nuts, peanuts and sesame seeds are oily seeds (a nut is just a big seed really).

2) These oils can be extracted and used for food or for fuel.

3) To get the oil out, the plant material is crushed. The next step is to press the crushed plant material between metal plates and squash the oil out. This is the traditional method of producing olive oil.

4) Oil can be separated from crushed plant material by a centrifuge — rather like using a spin-dryer to get water out of wet clothes. Or solvents can be used to get the oil from the plant material.

5) Distillation is used to refine the oil, and it also removes water, solvents and impurities.

Vegetable oils are used in food

1) Vegetable oils provide a lot of energy.

2) There are other nutrients in vegetable oils too — for example, oils from seeds contain vitamin E.

3) Vegetable oils contain essential fatty acids, which the body needs for many metabolic processes.

4) Vegetable oils tend to be unsaturated, while animal fats tend to be saturated.

5) In general, saturated fats are less healthy than unsaturated fats (as saturated fats increase the amount of cholesterol in the blood, which can block up the arteries and increase the risk of heart disease).

Plant oils can be good for you

Before fancy stuff from abroad like olive oil, we fried our bacon and eggs in lard. Vegetable oils, like olive oil, are better for use than animal fats, like lard, because they're usually unsaturated. Too much lard in your diet could increase your risk of getting heart disease.

Plant Oils in Food

Oils and fats can be saturated or unsaturated

- Oils and fats contain long-chain molecules with lots of carbon atoms.
- (Unsaturated oils contain double bonds between some of the carbon atoms in their carbon chains.)
  - Monounsaturated fats contain one C=C double bond somewhere in their carbon chains.
  - Polyunsaturated fats contain more than one C=C double bond.
- You can use the 'bromine water' test to check whether an oil or fat is saturated — see page 125.

Unsaturated oils can be hydrogenated

1) Unsaturated vegetable oils are liquid at room temperature.

2) They can be hardened by reacting them with hydrogen in the presence of a nickel catalyst at about 60 °C. This is called hydrogenation. The hydrogen reacts with the double-bonded carbons and opens out the double bonds.

3) Hydrogenated oils have higher melting points than unsaturated oils, so they're more solid at room temperature. This makes them useful as spreads and for baking cakes.

4) Margarine is usually made from partially hydrogenated vegetable oil — turning all the double bonds in vegetable oil to single bonds would make margarine too hard and difficult to spread. Hydrogenating most of them gives margarine a nice, buttery, spreadable consistency.

5) But partially hydrogenating vegetable oils means you end up with a lot of so-called trans fats. And there's evidence to suggest that trans fats are very bad for you.

You need to know the difference between saturated and unsaturated

This is tricky stuff. In a nutshell... there's saturated and unsaturated fats, which are generally bad and good for you (in that order) — easy enough. But... partially hydrogenated vegetable oil (which is unsaturated) is bad for you. Too much of the wrong types of fats can lead to heart disease. Got that...

Warm-Up and Exam Questions

Warm-up Questions

1) How can cooking food make it safer for humans to eat?
2) What effect does cooking eggs have on the proteins in them?
3) What are antioxidants?
4) Describe the process used to extract and refine oil from plants.
5) What is the difference between saturated fats and unsaturated oils?
6) What conditions are used for the hydrogenation of unsaturated vegetable oils?
7) Give an advantage of using artificial sweeteners instead of sugar in processed foods.

Exam Questions

1   Cooking potatoes breaks down

    A   cellulite

    B   protein

    C   cellulose

    D   celluloid

(1 mark)

2   Food colourings can be identified using chromatography. Outline a method
you could use to identify the different colourings in a food sample.

(3 marks)

3   Mayonnaise contains emulsifiers — molecules that have a hydrophilic end and
a hydrophobic end.

   (a)   On this diagram of an emulsifier molecule, label the hydrophilic and hydrophobic ends.

(1 mark)

   (b)   Explain the meanings of hydrophilic and hydrophobic.

(2 marks)

   (c)   The diagram below shows an oil molecule in water.
Show on the diagram how emulsifier molecules arrange themselves.

(1 mark)

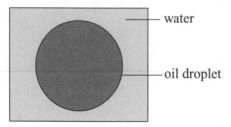

   (d)   What effect do emulsifiers have on the mayonnaise?

(1 mark)

Plant Oils as Fuel

Fuel from vegetable oil is possible too — but as always, you have to weigh up the pros and cons.

Vegetable oils can be used to produce fuels

1) Vegetable oils such as rapeseed oil and soybean oil can be processed and turned into fuels.

2) Vegetable oil provides a lot of energy — that's why it's suitable for use as a fuel.

3) A particularly useful fuel made from vegetable oils is called biodiesel. Biodiesel has similar properties to ordinary diesel fuel — it burns in the same way, so you can use it to fuel a diesel engine.

4) Most diesel engines can burn 100% biodiesel, but usually biodiesel is mixed with ordinary diesel.

5) Engines burning biodiesel produce 90% as much power as engines burning ordinary diesel.

Biodiesel is a renewable fuel

1) Biodiesel comes from plant crops, which can be planted and harvested every year. You can always keep making biodiesel.

2) Compare this to ordinary diesel, which is made by distilling crude oil. Crude oil was formed millions of years ago and it'll take millions of years to make more — once it runs low that's it.

Biodiesel releases less pollution than ordinary diesel

1) Engines burning biodiesel produce much less sulphur dioxide pollution than engines burning diesel or petrol.

2) Burning biodiesel doesn't release as many "particulates" as burning diesel or petrol.

Particulates are little pieces of solid crud that you get in smoke and car exhausts.

3) Biodiesel is also biodegradable and it's less toxic than regular diesel.

4) Biodiesel engines do release the same amount of carbon dioxide (CO_2) as ordinary diesel engines. BUT biodiesel comes from recently grown plants. The plants took in carbon dioxide from the air when they were alive, and it's this same carbon which is released again when the biodiesel is burned. So net increase in carbon dioxide in the atmosphere: nil.

5) Regular diesel, on the other hand, comes from crude oil, which has been under the ground for millions of years. The carbon in crude oil was taken out of the atmosphere millions of years ago. Burning regular diesel does create a net increase in carbon dioxide in the atmosphere.

Biodiesel is expensive and it's difficult to make enough

1) We can't make enough biodiesel to replace regular diesel — there aren't enough veg oil crops. Biodiesel can be made from used vegetable oil, but there isn't enough of that either.

2) Because of this, biodiesel is expensive. Most people won't want to use it until it's cheaper.

3) Biodiesel has fewer drawbacks than some other "green" car fuels like biogas or electricity, though. Car engines need modification to run on gas — most diesel cars run on biodiesel without any tinkering. And biodiesel could use the same filling stations and pumps as diesel. (Compare this with electric cars, which would need a new network of recharging stations.)

Ethanol

There are different kinds of alcohol, but the one that's in beer, wine and so on is ethanol.

Ethanol can be made by fermentation

1) <u>Fermentation</u> is the process of using <u>yeast</u> to convert <u>sugars</u> into <u>ethanol</u>. Carbon dioxide is also produced. Here's the equation:

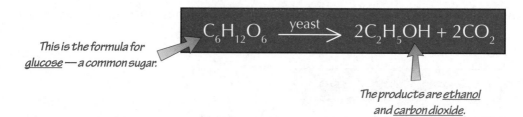

This is the formula for <u>*glucose*</u> *— a common sugar.*

$$C_6H_{12}O_6 \xrightarrow{\text{yeast}} 2C_2H_5OH + 2CO_2$$

The products are <u>*ethanol*</u> *and* <u>*carbon dioxide*</u>.

2) The yeast cells contain an <u>enzyme</u> called <u>zymase</u>. (Enzymes are naturally occurring <u>catalysts</u>.)

3) Fermentation happens fastest at a temperature of about 30 °C. At lower temperatures, the reaction slows down. If it's <u>too hot</u> the enzyme in the yeast is <u>destroyed</u>.

4) It's important to prevent <u>oxygen</u> getting to the fermentation process. If oxygen is present, a different reaction happens and you don't get ethanol.

5) When the <u>concentration</u> of alcohol reaches about 10 to 20%, the fermentation reaction <u>stops</u>, because the yeast gets <u>killed off</u> by the alcohol.

6) Different types of alcoholic drinks are made using sugars from different sources — usually from grains, fruits or vegetables, e.g. <u>barley</u> is used to make <u>beer</u> and grapes are used for making <u>wine</u>.

7) The fermented mixture can be <u>distilled</u> to produce more concentrated alcohol. Brandy is distilled from wine, whisky is distilled from fermented grain and vodka's distilled from fermented grain or potatoes.

8) The ethanol produced this way can also be used as quite a cheap <u>fuel</u> in countries which don't have oil reserves for making <u>petrol</u> (see next page).

Ethanol

There's more than one way to make ethanol...

Ethene can be reacted with steam to produce ethanol

1) Ethene (C_2H_4) will react with steam (H_2O) to make ethanol.

$$H_2C=CH_2 \ + \ H_2O \longrightarrow H-\underset{\underset{H}{|}}{\overset{\overset{H}{|}}{C}}-\underset{\underset{H}{|}}{\overset{\overset{H}{|}}{C}}-O-H$$

2) The reaction needs a temperature of 300 °C and a pressure of 70 atmospheres.

3) Phosphoric acid is used as a catalyst.

4) At the moment this is a cheap process, because ethene's fairly cheap and not much of it is wasted.

5) The trouble is that ethene's produced from crude oil, which is a non-renewable resource and which will start running out fairly soon. This means using ethene to make ethanol will become very expensive.

Alcohol can be used as a fuel

1) Ethanol can be used as fuel. It burns to give just CO_2 and water.

2) Cars can be adapted to run on a mixture of about 10% ethanol and 90% petrol — 'gasohol'. Some countries (e.g. Brazil) make extensive use of gasohol. It's best used in areas where there's plenty of fertile land for growing the crops needed, and good crop-growing weather.

3) Using gasohol instead of pure petrol means that less crude oil is being used. Another advantage is the crops needed for ethanol production absorb CO_2 from the atmosphere in photosynthesis while growing. This goes some way towards balancing out the release of CO_2 when the gasohol is burnt.

4) But distilling the ethanol after fermentation needs a lot of energy, so it's not a perfect solution.

OH — it's all about ethanol

Make sure you learn the different ways that ethanol can be produced, that means you need to learn those equations as well, it'll also be useful if you can remember a few uses of alcohol too.

Warm-Up and Exam Questions

The exam questions are pretty much like those you'll get in the real exam. Don't be tempted to look the answers up in the back of the book — they won't be there in the real thing.

Warm-up Questions

1) What is biodiesel? Give one advantage of biodiesel compared to normal diesel.
2) What useful fuel can be produced by fermentation?
3) Fermentation involves the substance zymase. What is zymase?
4) What is gasohol? Name a country where it is used extensively.
5) Whisky, brandy and vodka all involve distillation in their production.
   What effect does the distillation process have?

Exam Questions

1   Match the words **A**, **B**, **C** and **D** with numbers **1 – 4** in the sentences.

   A   oxygen

   B   sugar

   C   carbon dioxide

   D   glucose

   The raw material for fermentation is ...**1**..., a type of ...**2**....

   Fermentation produces the gas ...**3**... as well as alcohol.

   However if ...**4**... is present a different reaction happens.

(4 marks)

2   (a)   Ethanol can be synthesised in the following reaction:

   Reactant A is derived from crude oil. Explain why the cost of making ethanol by the synthesis method above is likely to increase.

(1 mark)

   (b)   Ethanol burns in oxygen to give carbon dioxide and water.

   (i)   Write a balanced symbol equation for this reaction.

(2 marks)

   (ii)   One method of ethanol production partially balances out the carbon dioxide produced when ethanol burns. Describe this method and explain why it helps balance out the carbon dioxide emissions from burning the ethanol.

(2 marks)

Hydration and Dehydration

There are loads of different kinds of chemical reaction — for example, hydration, dehydration and thermal decomposition. There are many others too, but you need to get your head round these ones for sure.

Hydration reactions have **water** as a **reactant**

1) Water's dead useful stuff. It's used as a solvent because it dissolves so many substances.

2) Water's also used in hydration reactions.

3) Hydration reactions are ones where water reacts with another substance to form a new product.

> Example: Calcium oxide (quicklime) reacts with water to produce calcium hydroxide (slaked lime).
>
> **calcium oxide + water → calcium hydroxide**
>
> **$CaO + H_2O → Ca(OH)_2$**

> Example: Ethene (C_2H_4) will react with steam (H_2O) to make ethanol.
>
> **ethene + steam → ethanol**
>
> **$C_2H_4 + H_2O → C_2H_5OH$**

Dehydration reactions have **water** as a **product**

Dehydration reactions are where water is removed from one or more substances, forming new products.

> Example: Dehydration reactions separate carbohydrates into carbon and water. Concentrated sulphuric acid is a strong dehydrating agent, and it'll grab water from other compounds. It acts as a catalyst to convert sucrose (sugar) into water and a spongy lump of black carbon.
>
> **sucrose** $\xrightarrow{\text{catalyst of conc. sulphuric acid}}$ **carbon + water**
>
> *Your teacher might demonstrate this in the fume cupboard.*

Reactions which join small molecules together to make a bigger molecule sometimes produce a molecule of water as well. These are usually called condensation reactions, but they're a kind of dehydration reaction too.

> Example: Alcohol reacts with organic acids to form an ester and water.
>
>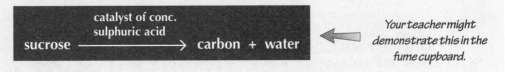
>
> **ethanol + ethanoic acid → ethyl ethanoate + water**

Thermal Decomposition

Carbonates and hydrogencarbonates release CO₂ when heated

1) Carbonates and hydrogencarbonates release <u>carbon dioxide gas</u> (CO_2) when they're heated.

2) It's an example of <u>thermal decomposition</u>, which is when a substance <u>breaks down</u> into simpler substances <u>when heated</u>.

3) <u>Learn the word equations</u> for the thermal decomposition of carbonates and hydrogencarbonates:

> **calcium carbonate → calcium oxide + carbon dioxide**

> **sodium hydrogencarbonate → sodium carbonate + carbon dioxide + water**

4) You can check it really is carbon dioxide that's released by testing it with <u>limewater</u> — CO_2 turns limewater <u>cloudy</u> when it's bubbled through.

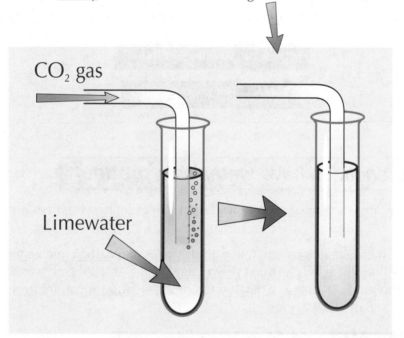

CO₂ gas

Limewater

5) Baking powder contains <u>sodium hydrogencarbonate</u>. Baking powder is added to cake mixture — the <u>carbon dioxide</u> produced when it's heated in the oven makes the cake <u>rise</u>.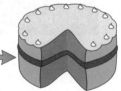

> *There's actually more to the action of <u>baking powder</u> than thermal decomposition. Baking powder is sodium hydrogencarbonate and an acid (usually 'cream of tartar'), which react to make carbon dioxide. Baking soda is plain sodium hydrogencarbonate (aka bicarbonate of soda).*

Thermal decomposition — when your vest unravels

Dehydration reactions take away hydrogen and oxygen from a molecule and produce <u>new products</u> — dehydrating sugar doesn't give you dry sugar, it gives you carbon.

Neutralisation and Oxidation

Have a look at this stuff on acids and bases.

Substances can be acids, bases or neutral

There's a sliding scale from very strong acid to very strong base, with neutral water in the middle.

These are the colours you get when you add universal indicator to an acid or an alkali.

An alkali is a soluble base.

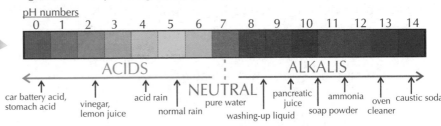

pH numbers
0 1 2 3 4 5 6 7 8 9 10 11 12 13 14

ACIDS   NEUTRAL   ALKALIS

car battery acid, stomach acid
vinegar, lemon juice
acid rain
normal rain
pure water
pancreatic juice
washing-up liquid
ammonia
soap powder
oven cleaner
caustic soda

Metal oxides and metal hydroxides are often bases

1) An acid and a base react together to form a salt and water. The products of the reaction aren't strongly acidic or alkaline — they're neutral. So it's called a neutralisation reaction. This is the equation for any neutralisation reaction:

$$acid + base \rightarrow salt + water$$

2) Metal oxides and metal hydroxides are generally bases. This means they'll react with acids to form a salt and water.

$$acid + metal\ oxide \rightarrow salt + water$$

$$acid + metal\ hydroxide \rightarrow salt + water$$

The combination of metal and acid decides the salt

Here are a couple of examples — an acid with: (i) a metal oxide, and (ii) a metal hydroxide...

hydrochloric acid   +   copper oxide $\rightarrow$ copper chloride   +   water

$$2HCl \quad + \quad CuO \quad \rightarrow \quad CuCl_2 \quad + \quad H_2O$$

sulphuric acid   +   calcium hydroxide $\rightarrow$ calcium sulphate   +   water

$$H_2SO_4 \quad + \quad Ca(OH)_2 \quad \rightarrow \quad CaSO_4 \quad + \quad 2H_2O$$

Rust is an example of a metal oxide

The word "rust" is only used for the corrosion of iron, not other metals.

Iron corrodes easily. In other words, it rusts.

1) When iron rusts, it's combining with oxygen (and also water). The iron gains oxygen to form iron(III) oxide. Water then becomes loosely bonded to the iron(III) oxide and the result is hydrated iron(III) oxide — which we call rust.

$$iron + oxygen + water \rightarrow hydrated\ iron(III)\ oxide$$

2) Combining with oxygen is an oxidation reaction. (The opposite process — removing oxygen from a compound — is called reduction. There's more about reduction on page 111.)

3) Unfortunately, rust is a soft crumbly solid that soon flakes off to leave more iron available to rust. And if the water's salty or acidic, rusting will take place a lot quicker. Cars in coastal places rust a lot because they get covered in salty sea spray. Cars in dry deserty places hardly rust at all.

Chemical Tests

If you have a mystery substance, there are various chemical tests you can do to find out what it is.

Flame tests — spot the colour

1) Some metals give a characteristic <u>colour</u> when heated, as you see every November 5th when a <u>firework explodes</u>. So, remember, remember...

> Potassium, K, burns with a lilac flame.

To flame-test a compound, dip a <u>clean wire loop</u> into a sample of the compound, and put the wire loop in the clear blue part of the Bunsen flame (the hottest bit).

> Sodium, Na, burns with a yellow/orange flame.

> Calcium, Ca, burns with a brick-red flame.

> Copper, Cu, burns with a blue-green flame.

2) So if you stick a bit of <u>copper wire</u> in a Bunsen flame, you'll see a <u>blue-green flame</u>.

3) But flame tests don't just work when you've got a sample of a <u>pure element</u> — they also work with a <u>compound</u> that contains that element. So if you stick a sample of, say, <u>copper sulphate</u> in a Bunsen flame, you'll also see a <u>blue-green flame</u>. Marvellous.

4) So, for example, a <u>forensic scientist</u> might examine a sample of <u>white crystals</u> found at a <u>crime scene</u>. The scientist could perform a <u>flame test</u> on the crystals and see that they burn with a <u>lilac coloured flame</u>. This test would show that the crystals must contain <u>potassium</u>.

Add sodium hydroxide and look for a coloured precipitate

1) You can sometimes find out what metal an <u>unknown substance</u> (e.g. a salt) contains by adding it (in solution) to <u>sodium hydroxide solution</u>.

2) When you do this a <u>chemical reaction</u> occurs and a <u>solid substance</u> containing the metal is formed (this solid substance is called a <u>precipitate</u>).

3) The precipitate has a <u>characteristic colour</u> which depends on the <u>metal</u> it contains.

4) Here are some typical precipitate colours...

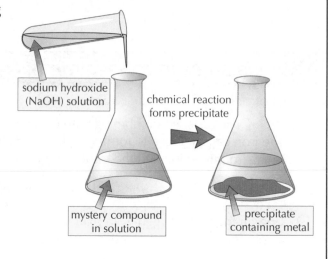

sodium hydroxide (NaOH) solution

chemical reaction forms precipitate

mystery compound in solution

precipitate containing metal

Transition metal	Colour of precipitate
Copper	Blue
Iron	Sludgy red or sludgy green
Zinc	White

Chemical Tests

You need to know the tests for common gases as well as common metals...

There are **tests** for **five common gases**

1) Chlorine

Chlorine <u>bleaches</u> damp <u>litmus paper</u>, turning it white. (It may turn <u>red</u> for a moment first though — that's because a solution of chlorine is <u>acidic</u>.)

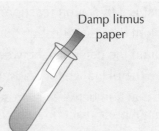
Damp litmus paper

2) Oxygen

Oxygen <u>relights</u> a <u>glowing splint</u>.

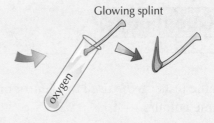
Glowing splint

3) Carbon dioxide

Carbon dioxide <u>turns limewater cloudy</u> — just bubble the gas through a test tube of limewater and watch what happens.

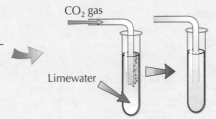
CO_2 gas
Limewater

4) Hydrogen

Hydrogen makes a "<u>squeaky pop</u>" with a <u>lighted splint</u> (as the hydrogen burns, forming water).

Squeaky pop!
Squeaky pop!

5) Ammonia

Ammonia <u>turns</u> damp <u>red litmus paper blue</u> (and has a very strong <u>smell</u>).

Get the colours right and you're halfway there

The principles are easy enough, but it's learning all those colours that's a bit of a pain in the neck.

Warm-Up and Exam Questions

Get stuck into these practice questions — it's the best way to check what you've learnt.
After all, answering questions is exactly what you'll have to do on the big day.

Warm-up Questions

1) What is a hydration reaction?
2) Explain why esterification can be thought of as a dehydration reaction.
3) Explain the difference between an alkali and a base.
4) Give the pH range for acids, the pH range for alkalis, and the pH number for neutral.
5) The reaction between sucrose and concentrated sulphuric acid produces water. What type of reaction is this?
6) What is the word equation for rust formation?

Exam Questions

1   Flame tests can be used to identify metals. Match up the metals **A, B, C** and **D** with the flame colours **1 - 4**.

A   sodium

B   potassium

C   calcium

D   copper

	Flame colour
1	lilac
2	yellow/orange
3	blue-green
4	brick-red

(4 marks)

2   (a)   Sodium hydroxide solution can be used to identify metals in an unknown substance. Briefly explain how.

(3 marks)

(b)   Give the tests (with results) for each of these gases:

(i)   oxygen

(ii)   ammonia

(iii)   hydrogen

(iv)   chlorine

(4 marks)

Exam Questions

3   Which of the following reactions is a neutralisation reaction?

   A $2Na + 2H_2O \rightarrow 2NaOH + H_2$

   B $Cu(OH)_2 + H_2SO_4 \rightarrow CuSO_4 + 2H_2O$

   C $CuSO_4 + 5H_2O \rightarrow CuSO_4.5H_2O$

   D $2NaHCO_3 \rightarrow CO_2 + Na_2CO_3 + H_2O$

(1 mark)

4   Rusting is accelerated by

   A   dry conditions

   B   salty, acidic conditions

   C   alkaline conditions

   D   grease and oil

(1 mark)

5   Which of the following is a base?

   A   copper nitrate

   B   copper oxide

   C   copper chloride

   D   copper sulphate

(1 mark)

6   When carbonates and hydrogencarbonates are heated they break down into simpler substances.

   (a)   What is the name of this process?

(1 mark)

   (b)   Sodium hydrogencarbonate breaks down on heating to form carbon dioxide and two other products.

      (i)   Give the formula for sodium hydrogencarbonate.

(1 mark)

      (ii)   Write the word equation for this reaction.

(1 mark)

      (iii) Describe a test you could do to show that carbon dioxide is produced in the reaction, and give the result you would expect.

(2 marks)

      (iv) Which household substance contains sodium hydrogencarbonate as an active ingredient and why?

(2 marks)

Energy Transfer in Reactions

Chemical reactions can either <u>release</u> heat energy, or <u>take in</u> heat energy.

Combustion is an **exothermic** reaction — heat's **given out**

An <u>EXOTHERMIC reaction</u> is one which <u>gives out energy</u> to the surroundings, usually in the form of <u>heat</u>, which is shown by a <u>rise in temperature</u>.

The best example of an <u>exothermic</u> reaction is <u>burning fuels</u>.
This obviously <u>gives out a lot of heat</u> — it's very exothermic.

In an **endothermic** reaction, heat is **taken in**

An <u>ENDOTHERMIC reaction</u> is one which <u>takes in energy</u> from the surroundings, usually in the form of <u>heat</u>, which is shown by a <u>fall in temperature</u>.

Endothermic reactions are <u>less common</u> and less easy to spot.
One example is <u>thermal decomposition</u>. Heat must be supplied to cause the compound to <u>decompose</u> (see page 178, or cracking on page 127).

Temperature changes help decide if a reaction's **exo** or **endo**

1) You can measure the amount of <u>energy produced</u> by a <u>chemical reaction</u> (in solution) by taking the <u>temperature of the reactants</u>, <u>mixing</u> them in a <u>polystyrene cup</u> and measuring the <u>temperature of the solution</u> at the <u>end</u> of the reaction. Easy.

2) Adding an <u>acid to an alkali</u> is an <u>exothermic</u> reaction. Measure the temperature of the alkali before you add the acid, then measure the temperature again after adding the acid and mixing — you'll see an <u>increase in temperature</u>.

3) Dissolving <u>ammonium nitrate</u> in water is an endothermic reaction. Adding a couple of spatulas of ammonium nitrate to a polystyrene cup of water results in a <u>drop in temperature</u>.

Here's the apparatus you'd use for these kinds of experiment:

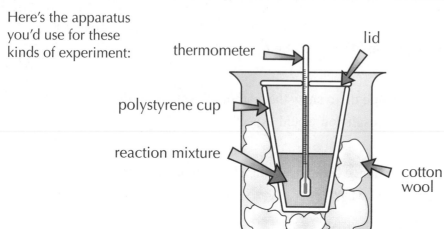

thermometer   lid
polystyrene cup
reaction mixture
cotton wool

Energy Transfer in Reactions

Energy must always be **supplied** to break bonds...

1) During a chemical reaction, <u>old bonds are broken</u> and <u>new bonds are formed</u>.

2) Energy must be <u>supplied</u> to break <u>existing bonds</u> — so bond breaking is an <u>endothermic</u> process.

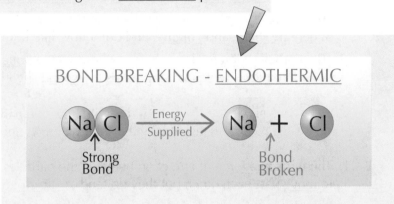

BOND BREAKING - <u>ENDOTHERMIC</u>

Energy Supplied

Strong Bond

Bond Broken

...and energy is always **released** when **bonds form**

1) Energy is <u>released</u> when new bonds are <u>formed</u> — so bond formation is an <u>exothermic</u> process.

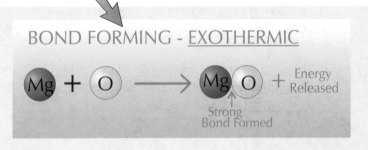

BOND FORMING - <u>EXOTHERMIC</u>

Energy Released

Strong Bond Formed

2) In an <u>exothermic</u> reaction, the energy <u>released</u> in bond formation is <u>greater</u> than the energy used in <u>breaking</u> old bonds.

3) In an <u>endothermic</u> reaction, the energy <u>required</u> to break old bonds is <u>greater</u> than the energy <u>released</u> when <u>new bonds</u> are formed.

Chemistry in "real-world application" shocker...

When you see <u>Stevie Gerrard</u> hobble off the pitch and press a bag to his leg, he's using an <u>endothermic reaction</u>. The cold pack contains an inner bag full of water and an outer one full of ammonium nitrate. When he presses the pack the inner bag <u>breaks</u> and they <u>mix together</u>. The ammonium nitrate dissolves in the water and, as this is an endothermic reaction, it <u>draws in heat</u> from Stevie's injured leg.

Kinetic Theory & Forces Between Particles

You can explain a lot of things (including perfumes) if you get your head round this lot.

States of matter — depend on the forces between particles

All stuff is made of particles (molecules, ions or atoms) that are constantly moving, and the forces between these particles can be weak or strong, depending on whether it's a solid, liquid or a gas.

Solids

1) There are strong forces of attraction between particles, which holds them in fixed positions in a very regular lattice arrangement.

2) The particles don't move from their positions, so all solids keep a definite shape and volume, and don't flow like liquids.

3) The particles vibrate about their positions — the hotter the solid becomes, the more they vibrate (causing solids to expand slightly when heated).

If you heat the solid (give the particles more energy), eventually the solid will melt and become liquid.

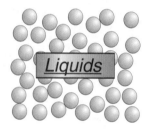

Liquids

1) There is some force of attraction between the particles. They're free to move past each other, but they do tend to stick together.

2) Liquids don't keep a definite shape and will flow to fill the bottom of a container.

3) The particles are constantly moving with random motion. The hotter the liquid gets, the faster they move. This causes liquids to expand slightly when heated.

If you now heat the liquid, eventually it will boil and become gas.

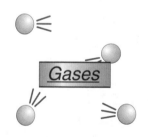

Gases

1) There's next to no force of attraction between the particles — they're free to move. They travel in straight lines and only interact when they collide.

2) Gases don't keep a definite shape or volume and will always fill any container. When particles bounce off the walls of a container they exert a pressure on the walls.

3) The particles move constantly with random motion. The hotter the gas gets, the faster they move. Gases either expand when heated, or their pressure increases.

How we smell stuff — volatility's the key

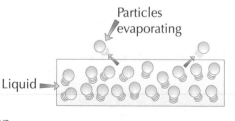

Particles evaporating

Liquid →

1) When a liquid is heated, the heat energy goes to the particles, which makes them move faster.

2) Some particles move faster than others.

3) Fast-moving particles at the surface will overcome the forces of attraction from the other particles and escape. This is evaporation.

4) How easily a liquid evaporates is called its volatility.

So... the evaporated particles are now drifting about in the air, the smell receptors in your nose pick up the chemical — and hey presto — you smell it.

Perfumes need to be quite volatile so they evaporate enough for you to smell them.

Chemical Reaction Rates

The <u>rate of a chemical reaction</u> is how fast the <u>reactants</u> are changed into <u>products</u> (the reaction is over when one of the reactants is completely used up).

Reactions can go at all sorts of **different rates**

1) One of the <u>slowest</u> is the <u>rusting</u> of iron.

2) Other slow reactions include <u>chemical weathering</u> — like acid rain damage to limestone buildings.

3) An example of a <u>moderate speed</u> reaction is a <u>metal</u> (e.g. magnesium) reacting with <u>acid</u> to produce a <u>gentle stream of bubbles</u>.

4) <u>Burning</u> is a <u>fast</u> reaction, but an <u>explosion</u> is <u>really fast</u> and releases a lot of gas. Explosive reactions are all over in a <u>fraction of a second</u>.

You can do an **experiment** to follow a **reaction**

The <u>rate of a reaction</u> that produces a gas can be observed by measuring how quickly the <u>gas is produced</u>. There are two ways of doing this:

1) Measure the change in mass

If you carry out the reaction on a balance, the <u>mass</u> will fall as the gas is released. You need to take <u>readings of the mass</u> at <u>regular time intervals</u>.

2) Measure the volume of gas given off

This method is pretty similar, except you use a <u>gas syringe</u> to measure the <u>volume of gas</u> given off after regular time intervals.

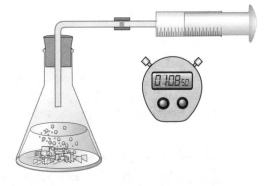

Whichever of these methods you use, you can plot your results on a <u>graph</u>. (See next page.)

Chemical Reaction Rates

*The **rate of a reaction** depends on **four things**:*

①   TEMPERATURE

②   CONCENTRATION — (or PRESSURE for gases)

③   SIZE OF PARTICLES — (or SURFACE AREA)

④   CATALYSTS (see page 189 for more)

The plot below shows how the speed of a particular reaction varies under different conditions. The quickest reaction is shown by the line that becomes flat in the least time. And the line that flattens out first must have the steepest slope — this makes it easy to spot the slowest and fastest reactions...

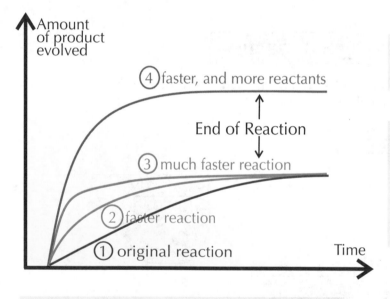

1) Graph 1 represents the original fairly slow reaction.

2) Graphs 2 and 3 represent the reaction taking place quicker but with the same initial amounts. (Notice that the slope of the graphs is steeper than for the original reaction.)

3) The increased rate could be due to any of these:

     a) increase in temperature
     b) increase in concentration (or pressure)
     c) catalyst added
     d) solid reactant crushed up into smaller bits.

4) Graph 4 produces more product as well as going faster. This can only happen if more reactant(s) are added at the start. Graphs 1, 2 and 3 all converge at the same level, showing that they all produce the same amount of product, although they take different times to get there.

Heat it up for a fast, furious reaction

First off... remember that the amount of product you get depends on the amount of reactants you start with. So all this stuff about the rate of a reaction is only talking about how quickly your products form — not how much of them you get. It's an important difference — so get your head round it asap.

Collision Theory

Reaction rates are explained perfectly by Collision Theory. It's really simple.

It just says that the rate of a reaction simply depends on how often and how hard the reacting particles collide with each other. The basic idea is that particles have to collide in order to react, and they have to collide hard enough as well.

More collisions **increases** the rate of reaction

All four methods of increasing the rate of reactions can be explained in terms of increasing the number of successful collisions between the reacting particles:

1) **Temperature** increases the number of collisions

When the temperature is increased the particles all move quicker. If they're moving quicker, they're going to have more collisions.

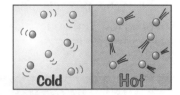

2) **Concentration** (or **pressure**) increases the number of collisions

If the solution is made more concentrated it means there are more particles of reactant knocking about between the water molecules, which makes collisions between the important particles more likely. In a gas, increasing the pressure means the molecules are more squashed up together so there are going to be more collisions.

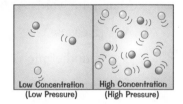

3) **Size of solid particles** (or **surface area**) increases collisions

If one of the reactants is a solid then breaking it up into smaller pieces will increase its surface area. This means the particles around it in the solution will have more area to work on so there'll be more useful collisions.

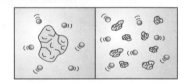

4) **Catalysts** increase the number of **successful** collisions

A catalyst is a substance that increases the rate of a reaction without being used up.

Some catalysts work by giving the reacting particles a surface to stick to where they can bump into each other. This increases the number of successful collisions (by lowering the activation energy) too.

Faster Collisions Also Increase the Rate of Reaction

Higher temperature increases the energy of the collisions, because it makes all the particles move faster.

Faster collisions are ONLY caused by increasing the temperature.

Reactions only happen if the particles collide with enough energy. At a higher temperature there will be more particles colliding with enough energy to make the reaction happen.

This initial energy is known as the activation energy, and it's needed to break the initial bonds. (See page 185.)

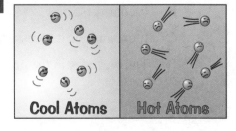

Warm-Up and Exam Questions

Same again — warm-up questions first, then exam questions.
Make the most of these pages by working through everything really carefully.

Warm-up Questions

1) Explain the terms 'exothermic' and 'endothermic'.
2) Give two examples of exothermic reactions.
3) Give two ways of following the rate of a reaction which produces a gas.
4) Finish this sentence: 'Collision Theory states that the rate of a reaction depends on…'
5) What is activation energy?
6) How can you make reactant particles move around faster? Explain your answer.

Exam Questions

1   Match the words **A**, **B**, **C** and **D** with numbers **1 – 4** in the sentences.

   A   releases

   B   endothermic

   C   requires

   D   exothermic

   Breaking bonds …**1**… energy and so is an …**2**… process.

   If a reaction …**3**… more energy than is taken in, the reaction is …**4**….

(4 marks)

2   Which of the following would reduce the rate of a chemical reaction?

   A   Reducing the surface area of the reactants.

   B   Increasing the temperature of the reaction.

   C   Raising the concentration of the reactants.

   D   Increasing the surface area of the reactants.

(1 mark)

3   Which of the following happens when a gas is heated?

   A   Its pressure decreases.

   B   It contracts.

   C   It expands or its pressure increases.

   D   The particles in the gas get bigger.

(1 mark)

Exam Questions

4 (a) Using the ideas of kinetic theory, explain:

   (i)   why liquids flow.

   (2 marks)

   (ii) why gases fill their containers.

   (2 marks)

   (iii) why solids have a fixed shape.

   (2 marks)

   (b) (i)   Explain the process of evaporation, in terms of particles.

   (1 mark)

   (ii) Explain why perfumes are made from substances which evaporate easily.

   (1 mark)

5 (a) The graph shows the amount of product produced over time in a reaction.

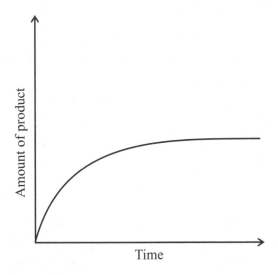

   (i)   Why does the graph level out?

   (1 mark)

   (ii) On the graph, mark the point where the reaction is fastest with an X.

   (1 mark)

   (iii) On the same set of axes, draw the curve you would expect if the reaction was
         repeated at a higher temperature, but with everything else staying the same.

   (2 marks)

   (b) Explain, in terms of collision theory, why increasing the pressure in a reaction
        between two gases increases the rate of the reaction.

   (2 marks)

Revision Summary for Section Six

The only way that you can tell if you've learned this module is to test yourself. Try these questions, and if there's something you don't know, it means you need to go back and learn it. Even if it is all that tricky business about rates of reaction. And don't miss any questions out — you don't get a choice about what comes up on the exam so you need to be sure that you've learnt it all.

1) What are polymers?
2) Plastic bags stretch and melt easily. Are the forces between the polymer chains weak or strong?
3) Explain how fabrics can be made both breathable and waterproof.
4) What is the name for the substance that gives a paint its colour?
5) Give three properties that a substance must have in order to make a good perfume.
6) Give two examples of new materials which were discovered by accident.
7) What kind of metal are "memory" glasses which remember their shape made from?
8) What are thermochromic pigments? How might these be used commercially?
9) What makes glow-in-the-dark watches glow in the dark?
10) Smart inks have been developed which get darker with time. The warmer they get, the faster they change colour. Explain why this is a useful property for a freshness indicator in food packaging.
11) Explain why nanoparticles are used in some sunscreens.
12) Explain why we don't eat uncooked potatoes.
13) Egg whites are clear and gloopy when raw, and white and rubbery when cooked. Name the kind of chemical in the eggs that is changed by heating.
14) Explain what is meant by "emulsion". List three foods which contain emulsions.
15) What is an E-number?
16) Give two advantages and two disadvantages of using food additives.
17) Describe how chromatography can be used to separate the different colours in a sweet.
18) Describe how olive oil is extracted from olives.
19) What kind of carbon-carbon bond do unsaturated oils contain?
20) Why are unsaturated oils hardened by reacting them with hydrogen?
21) State one advantage and one disadvantage of biodiesel compared to ordinary diesel.
22) Describe two ways in which ethanol can be made.
23) What products are formed by the thermal decomposition of calcium carbonate?
24) What do you get if an acid reacts with a base?
25) How does rust form? Is that an example of oxidation or reduction?
26) A forensic scientist carries out a flame test to identify a metal. The metal burns with a blue-green flame. a) Which metal does this result indicate? b) Describe a different test that the scientist could have used to identify the metal sample.
27) What's the test for each of the following: chlorine, hydrogen, oxygen, carbon dioxide, ammonia?
28) A substance keeps the same volume, but changes its shape according to the container it's held in. Is it a solid, a liquid or a gas? How strong are the forces of attraction between its particles? What does it mean if a liquid is said to be very volatile?
29)*A piece of magnesium is added to a dilute solution of hydrochloric acid, and hydrogen gas is produced. The experiment is repeated with a more concentrated hydrochloric acid. How can you tell from the experiment which concentration of acid produces a faster rate of reaction?
30) What four things affect the rate of a reaction?

* Answer on page 294.

Moving and Storing Heat

When it starts to get a bit nippy, on goes the heating to warm things up a bit. Heating is all about the transfer of energy. Here are a few useful definitions to begin with.

Heat is a measure of energy

1) When a substance is heated, its particles gain energy. This energy makes the particles in a gas or a liquid move around faster. In a solid, the particles vibrate more rapidly.

2) This energy is measured on an absolute scale. (This means it can't go lower than zero, because there's a limit to how slow particles can move.) The unit of heat energy is the joule (J).

Temperature is a measure of hotness

1) The hotter something is, the higher its temperature.

2) Temperature is usually measured in °C (degrees Celsius), but there are other temperature scales, like °F (degrees Fahrenheit).

Energy tends to flow from hot objects to cooler ones — e.g. warm radiators heat the cold air in your room. And the bigger the temperature difference, the faster heat is transferred. Kinda makes sense.

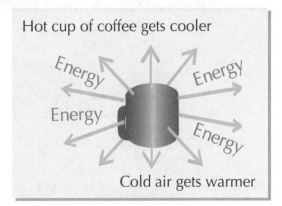

Hot cup of coffee gets cooler

Energy Energy Energy Energy

Cold air gets warmer

> If there's a DIFFERENCE IN TEMPERATURE between two places, then ENERGY WILL FLOW between them.

Specific heat capacity tells you how much energy stuff can store

1) It takes more heat energy to increase the temperature of some materials than others.

2) Materials which need to gain lots of energy to warm up also release loads of energy when they cool down again. They can 'store' a lot of heat.

E.g. you need 4200 J to warm 1 kg of water by 1 °C, but only 139 J to warm 1 kg of mercury by 1 °C.

3) The measure of how much energy a substance can store is called its specific heat capacity.

It's all about energy...

Heat is energy, heat is energy — get that in your head and this topic will really start to make sense. It also explains why heat flows from hot to cold. Particles are lazy — they don't want to be buzzing around with lots of energy, they'd much rather pass that energy on and take it easy. Chill...

Moving and Storing Heat

Specific heat capacity is different for different materials — which makes it a favourite calculation topic for examiners. Make sure you know the formula below so they don't catch you out.

You need to be able to calculate *specific heat capacity*

1) Specific heat capacity is the amount of energy needed to raise the temperature of 1 kg of a substance by 1 °C. Water has a specific heat capacity of 4200 J/kg/°C.

2) The specific heat capacity of water is high. Once water's heated, it stores a lot of energy, which makes it good for central heating systems. Also, water's a liquid so it can easily be pumped around a building.

3) You'll have to do calculations involving specific heat capacity.
   This is the equation to learn:

> **Energy = Mass × Specific Heat Capacity × Temperature Change**

> EXAMPLE: How much energy is needed to heat 2 kg of water from 10 °C to 100 °C?
>
> ANSWER: Energy needed = 2 × 4200 × 90 = 756 000 J

If you're not working out the energy, you'll have to rearrange the equation, so this formula triangle will come in dead handy.

You cover up the thing you're trying to find.
The parts of the formula you can still see are what it's equal to.

> EXAMPLE: An empty 200 g aluminium kettle cools down from 115 °C to 10 °C, losing 19 068 J of heat energy. What is the specific heat capacity of aluminium?
>
> ANSWER: SHC = $\dfrac{\text{Energy}}{\text{Mass} \times \text{Temp Ch}}$ = $\dfrac{19\ 068}{0.2 \times 105}$ = 908 J/kg/°C

Remember — you need to convert the mass to kilograms first.

I wish I had a high specific fact capacity...

So there are two reasons why water's used in central heating systems — it's a liquid and it has a high specific heat capacity. This makes water good for cooling systems too. Water can absorb a lot of energy and carry it away. Water-based cooling systems are used in car engines and some computers.

Melting and Boiling

If you've ever made a cup of tea you'll know you need energy to boil water — usually supplied by the kettle. But the energy doesn't just go into raising the water temperature...

You need to *put in energy* to *break intermolecular bonds*

1) When you heat a liquid, the <u>heat energy</u> makes the <u>particles move faster</u>. Eventually, when enough of the particles have enough energy to overcome their attraction to each other, big bubbles of <u>gas</u> form in the liquid — this is <u>boiling</u>.

2) It's similar when you heat a solid. <u>Heat energy</u> makes the <u>particles vibrate faster</u> until eventually the forces between them are overcome and the particles start to move around — this is <u>melting</u>.

Heating

When a substance is <u>melting</u> or <u>boiling</u>, you're still putting in <u>energy</u>, but the energy's used for <u>breaking intermolecular bonds</u> rather than raising the temperature — there are <u>flat spots</u> on the heating graph.

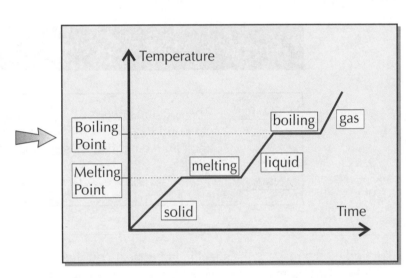

Cooling

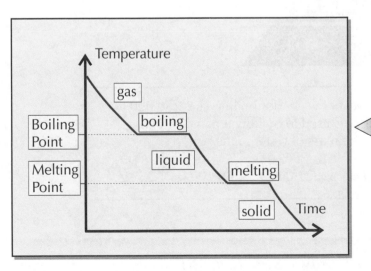

When a substance is <u>condensing</u> or <u>freezing</u>, bonds are <u>forming</u> between particles, which <u>releases</u> energy. This means the <u>temperature doesn't go down</u> until all the substance has turned into a liquid (condensing) or a solid (freezing).

Melting and Boiling

If you heat up a pan of water on the stove, the water never gets any hotter than 100 °C. You can <u>carry on heating it up</u>, but the <u>temperature won't rise</u>. It's all to do with <u>latent heat</u>...

Specific latent heat is the energy needed to change state

1) The <u>specific latent heat of melting</u> is the <u>amount of energy</u> needed to <u>melt 1 kg</u> of material <u>without changing its temperature</u> (i.e. the material's got to be at its melting temperature already).

2) The <u>specific latent heat of boiling</u> is the <u>energy</u> needed to <u>boil 1 kg</u> of material <u>without changing its temperature</u> (i.e. the material's got to be at its boiling temperature already).

3) Specific latent heat is <u>different</u> for <u>different materials</u>, and it's different for <u>boiling</u> and <u>melting</u>. You don't have to remember what all the numbers are, though. Phew.

4) There's a <u>formula</u> to help you with all the <u>calculations</u>. And here it is:

Energy = Mass × Specific Latent Heat

<u>EXAMPLE</u>: The specific latent heat of water (for melting) is 334 000 J/kg. How much energy is needed to melt an ice cube of mass 7 g at 0 °C?

<u>ANSWER</u>: Energy = 0.007 × 334 000 J = <u>2338 J</u>

If you're finding the mass or the specific latent heat you'll need to divide, not multiply — just to make your life a bit easier, here's the formula triangle.

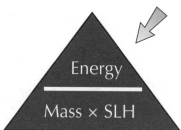

<u>EXAMPLE</u>: The specific latent heat of water (for boiling) is 2 260 000 J/kg. 2 825 000 J of energy is used to boil dry a pan of water at 100 °C. What was the mass of water in the pan?

<u>ANSWER</u>: Mass = Energy ÷ SLH = 2 825 000 ÷ 2 260 000 J = <u>1.25 kg</u>

It's all quite complicated but you really need to learn it

Melting a solid or boiling a liquid means you've got to <u>break bonds</u> between particles. That takes energy. Specific latent heat is just the amount of energy you need per kilogram of stuff. Incidentally, this is how <u>sweating</u> cools you down — your body heat's used to change liquid sweat into gas. Nice.

Warm-Up and Exam Questions

Take your time with these questions — and don't miss out the tricky-looking parts.
If any of them baffle you, it's not too late to take another peek over the section.

Warm-up Questions

1) What is heat?
2) Under what conditions will heat flow?
3) When a boiling liquid is heated, what is the heat energy used for?
4) Explain what is meant by 'specific latent heat of melting'.

Exam Questions

1   A 0.5 kg iron block is heated from 20 °C to 100 °C. This takes 18 000 J of energy.
    The specific heat capacity of iron is

    A   250 J/kg°C

    B   450 J/kg°C

    C   650 J/kg°C

    D   850 J/kg°C

    (1 mark)

2   The graph shows a cooling curve for a gas. Match the labels **A**, **B**, **C** and **D** with
    points **1- 4** on the graph.

    A   Cooling liquid

    B   Condensing

    C   Freezing

    D   Cooling gas

    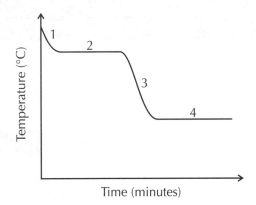

    (4 marks)

3   Sam is boiling some water. She uses a kettle which delivers 2000 J every second.
    Assume the kettle is 100% efficient.

    (a) How much heat is delivered to the water in two minutes?

    (2 marks)

    (b) The specific latent heat for boiling water is 2 260 000 J/kg. Once the water in the
         kettle is at 100 °C, what mass of water will turn to steam if it carries on boiling for
         two minutes?

    (2 marks)

Conduction

If you build a house, there are regulations about doing it properly, mainly so that it doesn't fall down, but also so that it <u>keeps the heat in</u>. Easier said than done — there are several ways that heat is 'lost'.

Conduction is most important in solids

Houses lose a lot of heat through their windows even when they're shut. One reason for this is that heat flows from the warm inside face of the window to the cold outside face by <u>conduction</u>.

1) In a <u>solid</u>, the particles are held tightly together. So when one particle <u>vibrates</u>, it <u>bumps into</u> other particles nearby and quickly passes the vibrations on.

2) Particles which vibrate <u>faster</u> than others pass on their <u>extra kinetic energy</u> (that's <u>movement</u> energy) to <u>neighbouring particles</u>. These particles then vibrate faster themselves.

3) This process continues throughout the solid and gradually the extra kinetic energy (or <u>heat</u>) is spread all the way through the solid. This causes a <u>rise in temperature</u> at the <u>other side</u>.

> <u>CONDUCTION OF HEAT</u> is the process where <u>vibrating particles</u> pass on <u>extra kinetic energy</u> to <u>neighbouring particles</u>.

4) <u>Metals</u> are really <u>good conductors of heat</u> — that's why they're used for <u>saucepans</u>.

   <u>Non-metals</u> are good for <u>insulating</u> things — e.g. for saucepan <u>handles</u>.

   Metals are good conductors because free electrons (page 110) help "carry" the heat through the metal.

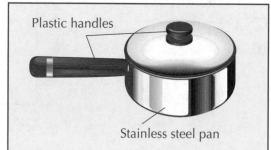

Plastic handles

Stainless steel pan

5) <u>Liquids and gases</u> conduct heat <u>more slowly</u> than solids — the particles aren't held so tightly together. So <u>air</u> is a good insulator.

It's all down to those free electrons

Conduction is like pass the parcel — each particle passes the heat on to its neighbour. They can't chuck heat across the room or anything like that — that's just not playing fair. Each particle vibrates — the more energy each has, the faster it vibrates — and passes energy onto its neighbours.

Convection

Convection occurs in *liquids* and *gases*

1) When you heat up a liquid or gas, the particles move faster, and the fluid (liquid or gas) <u>expands</u>, becoming <u>less dense</u>.

2) The <u>warmer</u>, <u>less dense</u> fluid <u>rises</u> above its <u>colder</u>, <u>denser</u> surroundings, like a hot air balloon does.

3) As the <u>warm</u> fluid <u>rises</u>, cooler fluid takes its place. As this process continues, you actually end up with a <u>circulation</u> of fluid (<u>convection currents</u>). This is how <u>immersion heaters</u> work.

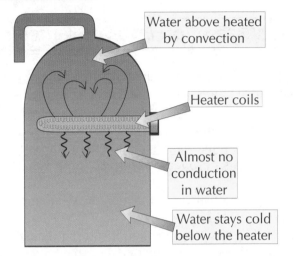

Water above heated by convection

Heater coils

Almost no conduction in water

Water stays cold below the heater

<u>CONVECTION</u> occurs when more energetic particles <u>move</u> from a <u>hotter region</u> to a <u>cooler region</u> — <u>and take their heat energy with them</u>.

4) <u>Radiators</u> in the home rely on convection to make the warm air <u>circulate</u> round the room.

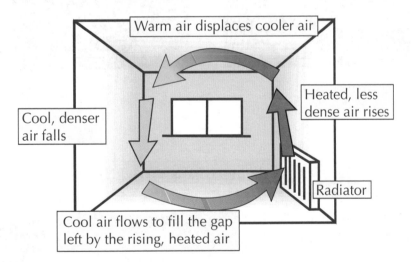

Warm air displaces cooler air

Cool, denser air falls

Heated, less dense air rises

Radiator

Cool air flows to fill the gap left by the rising, heated air

5) Convection <u>can't happen in solids</u> because the <u>particles can't move</u> — they just vibrate on the spot.

6) To <u>reduce convection</u>, you need to <u>stop the fluid moving</u>.

Clothes, blankets and cavity wall foam insulation all work by <u>trapping pockets of air</u>. The air can't move so the heat has to conduct <u>very slowly</u> through the pockets of air, as well as the material in between.

Remember: most 'radiators' don't radiate — they cause convection

If a <u>garden spade</u> is left outside in cold weather, the metal bit will always feel <u>colder</u> than the wooden handle. But it <u>isn't</u> colder — it just <u>conducts heat away</u> from your hand quicker. The opposite is true if the spade is left out in the sunshine — it'll <u>feel</u> hotter because it conducts heat into your hand quicker.

Heat Radiation

The other way heat can be transferred is by <u>radiation</u>.
This is very different from conduction and convection.

*Thermal **radiation** involves **emission** of **electromagnetic waves***

<u>Heat radiation</u> can also be called <u>infrared radiation</u>, and it consists purely of electromagnetic waves of a certain range of frequencies. It's next to visible light in the <u>electromagnetic spectrum</u> (see page 236).

1) <u>All objects</u> continually <u>emit</u> and <u>absorb</u> <u>heat radiation</u>. An object that's <u>hotter</u> than its surroundings <u>emits more radiation</u> than it <u>absorbs</u> (as it <u>cools</u> down). And an object that's <u>cooler</u> than its surroundings <u>absorbs more radiation</u> than it <u>emits</u> (as it <u>warms</u> up).

2) The <u>hotter</u> an object gets, the <u>more</u> heat radiation it <u>emits</u>.

3) You can <u>feel</u> this <u>heat radiation</u> if you stand near something <u>hot</u> like a fire or if you put your hand just above the bonnet of a recently parked car.

(recently parked car)

(after an hour or so)

Radiation is how we get heat from the Sun

1) <u>Radiation</u> can occur in a <u>vacuum</u>, like space. This is the <u>only way</u> that heat reaches us from the <u>Sun</u>.

2) Heat radiation only passes through substances that are <u>transparent</u> to infrared radiation — e.g. <u>air</u>, <u>glass</u> and <u>water</u>.

3) The <u>amount</u> of radiation emitted or absorbed by an object depends to a large extent on its <u>surface colour and texture</u>. This definitely <u>isn't true</u> for conduction and convection.

If radiation couldn't travel through a vacuum, we'd all be very cold

If you think about sunshine, it's easy to see that heat and light are similar things. They're both part of the electromagnetic spectrum and both travel in waves. The difference between them is their wavelength and frequency — but not their speed, all waves travel at the same speed.

Heat Radiation

The amount of heat **radiated** and **emitted** depends on...

1) Surface area

1) Heat is radiated from the surface of an object.

2) The bigger the surface area, the more waves can be emitted from the surface — so the quicker the transfer of heat.

3) This is why car and motorbike engines often have 'fins' — they increase the surface area so heat is radiated away quicker. So the engine cools quicker.

4) It's the same with heating something up — the bigger the surface area exposed to the heat radiation, the quicker it'll heat up.

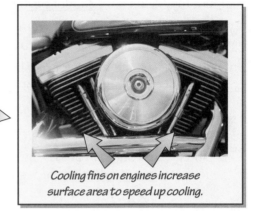

Cooling fins on engines increase surface area to speed up cooling.

2) Colour and texture

1) Matt black surfaces are very good absorbers and emitters of radiation. Painting a wood-burning stove matt black means it'll radiate as much heat as possible.

2) Light-coloured, smooth objects are very poor absorbers and emitters of radiation, but they effectively reflect heat radiation.

For example, some people put shiny foil behind their radiators to reflect radiation back into the room rather than heat up the walls.

Another good example is survival blankets for people rescued from snowy mountains — their shiny, smooth surface reflects the body heat back inside the blanket, and also minimises heat radiation being emitted by the blanket.

Revise heat radiation — well, at least absorb as much as you can

The most confusing thing about radiation is that those white things on your walls called 'radiators' actually transfer most of their heat by convection, as rising warm air. They do radiate some heat too, of course, but whoever chose the name 'radiator' obviously hadn't swotted up their physics first.

Saving Energy

Insulating *your home* **saves energy** *and* **money**

1) To save energy, you need to <u>insulate</u> your home. It <u>costs money</u> to buy and install the insulation, but it also <u>saves</u> you money, because your <u>heating bills</u> are lower.

2) Eventually, the <u>money you've saved</u> on heating bills will <u>equal</u> the <u>initial cost</u> of installing the insulation — the time this takes is called the <u>payback time</u>.

3) <u>Cheaper</u> methods of insulation are usually less effective — they tend to save you less money per year, but they often have <u>shorter payback times</u>.

4) If you <u>subtract</u> the <u>annual saving</u> from the <u>initial cost</u> repeatedly then <u>eventually</u> the one with the <u>biggest annual saving</u> must always come out as the winner, if you think about it.

5) But you might sell the house (or die) before that happens. If you look at it over, say, a <u>five-year period</u> then a cheap and cheerful <u>hot water tank jacket</u> wins over expensive <u>double glazing</u>.

Loft insulation

Fibreglass 'wool' laid across the loft floor reduces <u>conduction</u> through the ceiling into the roof space.
Initial Cost:　£200
Annual Saving: £100
Payback time: <u>2 years</u>

Hot water tank jacket

Lagging such as fibreglass wool reduces <u>conduction</u>.
Initial Cost:　£60
Annual Saving:　£15
Payback time:　<u>4 years</u>

Double glazing

Two layers of glass with an air gap between reduce <u>conduction</u>.
Initial Cost:　£2400
Annual Saving: £80
Payback time: <u>30 years</u>

Cavity walls & insulation

Two layers of bricks with a gap between them reduce <u>conduction</u>. <u>Insulating foam</u> is squirted into the gap between layers, trapping pockets of air to minimise <u>convection</u>.
Initial Cost:　£150
Annual Saving: £100
Payback time: <u>18 months</u>

The exact costs and savings will depend on the house — but these figures give you a rough idea.

Draught-proofing

Strips of foam and plastic around doors and windows stop hot air going out — reducing <u>convection</u>.
Initial Cost:　£100
Annual Saving:　£15
Payback time: <u>7 years</u>

Thick curtains

Reduce <u>conduction</u> and <u>radiation</u> through the windows.
Initial Cost:　£180
Annual Saving: £20
Payback time: <u>9 years</u>

Thermograms *show where your home is* **leaking heat**

A thermogram is a picture taken with a thermal imaging camera. Objects at <u>different temperatures</u> emit <u>infrared</u> rays of <u>different wavelengths</u>, which the thermogram displays as <u>different colours</u>.

In this thermogram, <u>red</u> shows where most heat is being lost. The houses on the left and right are losing bucket-loads of heat out of their roofs, but the one in the middle must have <u>loft insulation</u> as it's not losing half as much.

TONY MCCONNELL / SCIENCE PHOTO LIBRARY

Warm-Up and Exam Questions

Yes, you've got it — do the warm-up questions first, then when you think you're ready, have a go at the exam questions. If there's anything you can't do, make sure you go back and check on it.

Warm-up Questions

1) Describe the process of heat transfer by conduction.
2) Explain why heated air rises.
3) Heat radiation can also be called thermal radiation. What is another name for it?
4) Give two ways the nature of a surface could be changed so that the surface emits more heat radiation.
5) Give three methods of insulating a house.

Exam Questions

1   Warm air rises in the roof space of a house.
    What type of heat transfer is this?

    A   radiation

    B   conduction

    C   convection

    D   insulation

    (1 mark)

2   Mandy wants some blinds for her new conservatory so that it doesn't get too hot on very sunny days. Which ones should she choose?

    A   matt, black blinds

    B   matt, white blinds

    C   shiny, black blinds

    D   shiny, white blinds

    (1 mark)

3   The diagram shows the heat losses from Tom's house. Tom estimates that £300 of his annual heating bill is wasted on heat lost from the house.

    (a)   How much money does Tom waste every year in heat lost through the roof?

    (1 mark)

    (b)   Tom decides to insulate the loft. This costs him £350, but reduces the amount he spends on wasted heat to £255 per year.

    Calculate the payback time for fitting loft insulation in Tom's house.

    (2 marks)

25% through roof

35% through brick walls

15% through open windows and cracks in window frames

15% through concrete floor

10% through glass in windows

Energy

Learn these *nine types* of *energy*

You should know all of these <u>well enough</u> by now to list them <u>from memory</u>, including the examples:

1) <u>ELECTRICAL</u> Energy.................................... — whenever a <u>current</u> flows.
2) <u>LIGHT</u> Energy... — from the <u>Sun</u>, <u>light bulbs</u>, etc.
3) <u>SOUND</u> Energy.. — from <u>loudspeakers</u> or anything <u>noisy</u>.
4) <u>KINETIC</u> Energy, or <u>MOVEMENT</u> Energy..... — anything that's <u>moving</u> has it.
5) <u>NUCLEAR</u> Energy...................................... — released only from <u>nuclear reactions</u>.
6) <u>THERMAL</u> Energy or <u>HEAT</u> Energy.............. — <u>flows</u> from <u>hot objects</u> to colder ones.
7) <u>GRAVITATIONAL POTENTIAL</u> Energy........ — possessed by anything which can <u>fall</u>.
8) <u>ELASTIC POTENTIAL</u> Energy...................... — possessed by <u>springs</u>, <u>elastic</u>, <u>rubber bands</u>, etc.
9) <u>CHEMICAL</u> Energy..................................... — possessed by <u>foods</u>, <u>fuels</u>, <u>batteries</u> etc.

> The <u>last three</u> above are forms of <u>stored energy</u> because the energy is not obviously <u>doing</u> anything, it's kind of <u>waiting to happen</u>, i.e. waiting to be turned into one of the <u>other</u> forms.

There are two types of *"energy conservation"*

Try and get your head round the difference between these two:

1) "<u>ENERGY CONSERVATION</u>" is all about <u>using fewer resources</u> because of the damage they do and because they might <u>run out</u>. That's all <u>environmental stuff</u> — which is important to us, but fairly trivial on a <u>cosmic scale</u>.

2) The "<u>PRINCIPLE OF THE CONSERVATION OF ENERGY</u>", however, is one of the <u>major cornerstones</u> of modern physics. It's an <u>all-pervading principle</u> which governs the workings of the <u>entire physical Universe</u>. If this principle were not so, then life as we know it would simply cease to be.

The *principle of the conservation of energy* can be stated thus:

> <u>Energy</u> can never be <u>created nor destroyed</u>
> — it's only ever <u>converted</u> from one form to another.

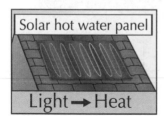

Solar hot water panel

Light → Heat

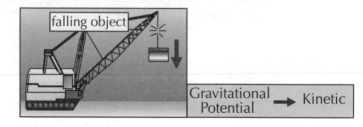

falling object

Gravitational Potential → Kinetic

Another <u>important principle</u> which you need to <u>learn</u> is this one:

> Energy is <u>only useful</u> when it can be <u>converted</u> from one form to another.

Efficiency

An open fire looks cosy, but a lot of its heat energy goes straight up the chimney, by convection, instead of heating up your living room. All this energy is 'wasted', so open fires aren't very efficient.

Machines always waste some energy

1) Useful machines are only useful because they convert energy from one form to another. Take cars for instance — you put in chemical energy (petrol or diesel) and the engine converts it into kinetic (movement) energy.

2) The total energy output is always the same as the energy input, but only some of the output energy is useful. So for every joule of chemical energy you put into your car you'll only get a fraction of it converted into useful kinetic energy.

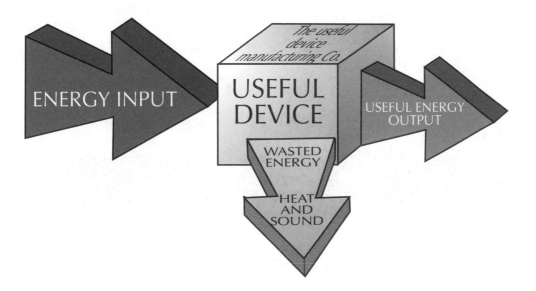

3) This is because some of the input energy is always lost or wasted, often as heat. In the car example, the rest of the chemical energy is converted (mostly) into heat and sound energy. This is wasted energy.

4) The less energy that is wasted, the more efficient the device is said to be.

The efficiency of a machine is defined as...

$$\text{Efficiency} = \frac{\text{USEFUL Energy OUTPUT}}{\text{TOTAL Energy OUTPUT}}$$

The input energy is ALWAYS more than the useful energy output

Efficiency is all about what goes in and how it comes out. Remember that conservation of energy thingy — what do you mean no? It's on the last page — anyway that means that however much energy you put in, you'll get that same amount out. The tricky bit is working out how much comes out usefully.

Efficiency

More *efficient* machines *waste less energy*

1) To work out the efficiency of a machine, first find out the Total Energy output.
   This is the same as the energy supplied to the machine — the energy input.

2) Then find how much useful energy the machine delivers — the USEFUL Energy output.
   The question might tell you this directly, or it might tell you how much energy is wasted.

3) Then just divide the smaller number by the bigger one to get a value for efficiency somewhere
   between 0 and 1. Easy. (If your number is bigger than 1, you've done the division upside down.)

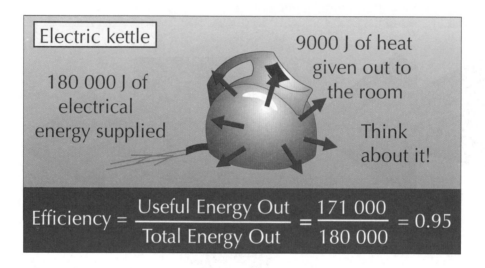

Electric kettle

180 000 J of electrical energy supplied

9000 J of heat given out to the room

Think about it!

$$\text{Efficiency} = \frac{\text{Useful Energy Out}}{\text{Total Energy Out}} = \frac{171\ 000}{180\ 000} = 0.95$$

4) You can convert the efficiency to a percentage, by multiplying it by 100. E.g. 0.6 = 60%.

5) In the exam you might be told the efficiency and asked to work out the total energy output,
   the useful energy output or the energy wasted. So you need to be able to rearrange the formula.

EXAMPLE: An ordinary light bulb is 5% efficient. If 1000 J of light
energy is given out, how much energy is wasted?

ANSWER: $\text{Total Output} = \dfrac{\text{Useful Output}}{\text{Efficiency}} = \dfrac{1000\ J}{0.05} = 20\ 000\ J,$

so Energy Wasted = 20 000 – 1000 = 19 000 J

Shockingly inefficient, those ordinary light bulbs. Low-energy light bulbs are roughly 4 times more efficient, and last about 8 times as long. They're more expensive though.

Efficiency questions are all more or less the same

Some new appliances (like washing machines and fridges) come with a sticker with a letter from A to H
on, to show how energy-efficient they are. A really well-insulated fridge might have an 'A' rating. But if
you put it right next to the oven, or never defrost it, it will run much less efficiently than it should.

Warm-Up and Exam Questions

It's no good learning all the facts in the world if you go to pieces and write nonsense in the exam. So you'd be wise to practise using all your knowledge to answer some questions.

Warm-up Questions

1) What type of energy is stored in food?
2) State the principle of the conservation of energy.
3) Modern appliances tend to be more energy efficient than older ones. What does this mean?
4) Give an example of a device that uses elastic potential energy.
5) Why is the efficiency of an appliance always less than 100%?

Exam Questions

1 A motor is supplied with 200 J of energy to lift a load.
The load gains 140 J of potential energy.
What is the efficiency of the motor?

   A   60 J

   B   70%

   C   0.3

   D   1.43

(1 mark)

2 What useful energy changes occur when a pop star sings into a microphone connected to an amplifier and loudspeaker? (Loudspeakers use electricity to vibrate a cone, which produces sound waves.)

   A   sound → kinetic → heat → sound

   B   chemical → sound → electrical → sound

   C   electrical → sound → heat → sound

   D   sound → electrical → kinetic → sound

(1 mark)

3 A hairdryer is supplied with 1200 J of electrical energy every second. The electrical energy is converted to 20 J of sound energy and 100 J of kinetic energy every second.

   (a) How much electrical energy does the hairdryer transform into heat energy every second?

(1 mark)

   (b) Suggest how the hairdryer could be made more efficient.

(1 mark)

Energy Sources

There are various different types of energy resource.
They fit into two broad types: renewable and non-renewable.

Non-renewable energy resources will run out one day

The non-renewables are the three FOSSIL FUELS and NUCLEAR:

1) Coal

2) Oil

3) Natural gas

4) Nuclear fuels (uranium and plutonium)

> a) They will all 'run out' one day.
> b) They all do damage to the environment.
> c) But they provide most of our energy.

There are environmental problems with non-renewables

1) All three fossil fuels (coal, oil and gas) release CO_2. For the same amount of energy produced, coal releases the most CO_2, followed by oil then gas. All this CO_2 adds to the greenhouse effect, and contributes to climate change. We could stop some of it entering the atmosphere — by 'capturing' it and burying it underground, for instance — but the technology is too expensive to be widely used yet.

2) Burning coal and oil releases sulphur dioxide, which causes acid rain.
   This is reduced by taking the sulphur out before it's burned, or cleaning up the emissions.

3) Coal mining makes a mess of the landscape, especially "open-cast mining".

4) Oil spillages cause serious environmental damage. We try to avoid them, but they'll always happen.

5) Nuclear power is clean but the nuclear waste is dangerous and difficult to dispose of (see page 211).

6) But non-renewable fuels are generally concentrated energy resources, reliable, and easy to use.

Learn about the non-renewables — before it's too late

There's lots more info about the various power sources over the next few pages. But the point is that none of them are ideal — they all have pros and cons, and the aim is to choose the least bad option overall. Unless we want to go without heating, light, transport, electricity... and so on. (I don't.)

Energy Sources

Renewable energy sources are often seen as the energy sources of the future — they're more friendly to the environment and won't run out like the non-renewables, but they do have problems of their own...

Renewable energy resources will *never run out*

The <u>renewables</u> are:

1) <u>Geothermal</u>
2) <u>Wind</u>
3) <u>Solar</u>
4) <u>Biomass</u>
5) <u>Waves</u>
6) <u>Tides</u>
7) <u>Hydroelectric</u>

> a) These will <u>never run out</u>.
> b) Most of them do <u>damage the environment</u>, but in <u>less nasty</u> ways than non-renewables.
> c) The trouble is they <u>don't provide much energy</u> and some of them are <u>unreliable</u> because they depend on the <u>weather</u>.

The Sun is the *ultimate source* of loads of *energy*

1) A lot of these energy sources can be traced back to the Sun.

2) <u>Every second</u> for the last few billion years or so, the Sun has been giving out <u>loads</u> of <u>energy</u> — mostly in the form of <u>heat</u> and <u>light</u>.

3) Some of that energy is <u>stored</u> here on Earth as <u>fossil fuels</u> (coal, oil and gas).

> Fossil fuels are the remains of plants and animals that lived long ago.

4) When we use wind power, we're also using the Sun's energy — the Sun heats the <u>air</u>, the <u>hot air rises</u>, cold air <u>whooshes in</u> to take its place (wind), and so on.

Nuclear, *geothermal* and *tidal* energy do *not* originate in the Sun

1) <u>Nuclear power</u> comes from the energy locked up in the <u>nuclei of atoms</u>.

2) Nuclear decay also creates heat inside the Earth for <u>geothermal energy</u>, though this happens much <u>slower</u> than in a nuclear reactor.

3) <u>Tides</u> are caused by the <u>gravitational attraction</u> of the Moon and Sun.

Most energy resources come from the Sun — but not all of them

The Sun really is rather useful, not only does it warm and light our little planet — it also provides lots of energy sources from which we can generate electricity. It's quite obvious that the Sun provides the energy for solar power, but don't forget about wind, waves and biomass they come from the Sun too.

Nuclear Energy

Well, who'd have thought... there's energy lurking about inside atoms.

Nuclear power uses uranium as fuel

1) A nuclear power station uses uranium to produce heat.

2) Nuclear power stations are expensive to build and maintain, and they take longer to start up than fossil fuel power stations. (Natural gas is the quickest.)

3) Processing the uranium before you use it causes pollution, and there's always a risk of leaks of radioactive material, or even a major catastrophe like at Chernobyl.

4) A big problem with nuclear power is the radioactive waste that you always get. And when they're too old and inefficient, nuclear power stations have to be decommissioned (shut down and made safe) — that's expensive too.

5) But there are many advantages to nuclear power. It doesn't produce any of the greenhouse gases which contribute to global warming. Also, there's still plenty of uranium left in the ground (although it can take a lot of money and energy to make it suitable for use in a reactor).

Nuclear fuel heats water to produce steam

1) The nuclear fuel produces heat.

2) The heat is taken to the steam generator using a liquid coolant.

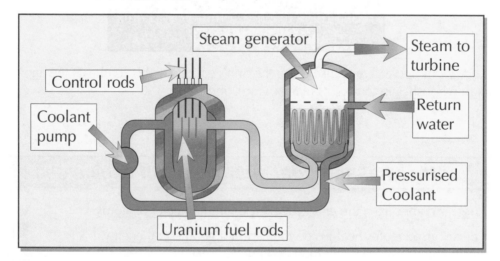

3) The water in the steam generator gets hot, and turns to steam.

4) The steam is used to turn a turbine and produce electricity.

Many people object to nuclear power because of the waste

Nuclear power sounds a bit scary at first, but the method of using it is just the same as for coal and oil and all that boring stuff — you make some heat, which boils water to make steam. I'm even thinking of getting a nuclear powered kettle to save time in the morning when I need that cup of coffee...

Nuclear and Geothermal Energy

Radioactive waste is difficult to dispose of safely

1) Most waste from nuclear power stations and hospitals is 'low-level' (only slightly radioactive). This kind of waste can be disposed of by burying it in secure landfill sites.

2) High-level waste is the really dangerous stuff — a lot of it stays highly radioactive for tens of thousands of years, and so has to be treated very carefully. It's often sealed into glass blocks, which are then sealed in metal canisters. These could then be buried deep underground.

3) However, it's difficult to find suitable places to bury high-level waste. The site has to be geologically stable (e.g. not suffer from earthquakes), since big movements in the rock could disturb the canisters and allow radioactive material to leak out. And even when geologists do find suitable sites, people who live nearby often object.

4) Not all radioactive waste has to be chucked out though — some of it is reprocessed. After reprocessing, you're left with more uranium (for reuse in power stations) and a bit of plutonium (which can be used to make nuclear weapons).

5) But nuclear power stations and reprocessing plants might be a target for terrorists — who could attack the plant, or use stolen material to make a 'dirty bomb'.

Geothermal energy — heat from underground

1) This is only possible in certain places where hot rocks lie quite near to the surface. The source of much of the heat is the slow decay of various radioactive elements including uranium deep inside the Earth.

2) Water is pumped in pipes down to hot rocks and returns as steam to drive a generator. This is actually brilliant free energy with no real environmental problems. The main drawback is the cost of drilling down several km.

3) Unfortunately there are very few places where this seems to be an economic option (for now).

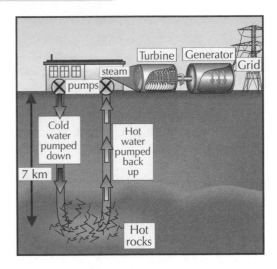

It might be expensive — but it'll last forever

Most of the UK's nuclear power stations are quite old, and will be shut down soon. There's a debate going on over whether we should build new ones. Some people say no — if we can't deal safely with the radioactive waste we've got now, we certainly shouldn't make lots more. Others say nuclear power is the only way to meet all our energy needs without causing catastrophic climate change.

Wind and Solar Energy

Wind and solar energy are both renewable — they'll never run out.

Wind farms — lots of little wind turbines

1) Wind power involves putting lots of wind turbines up in <u>exposed places</u> — like on <u>moors</u>, around the <u>coast</u> or <u>out at sea</u>.

2) Wind turbines convert the kinetic energy of moving air into electricity. The <u>wind</u> turns the <u>blades</u>, which turn a <u>generator</u>.

3) Wind turbines are quite cheap to run — they're very <u>tough</u> and reliable, and the wind is <u>free</u>.

4) Even better, wind power doesn't produce any <u>polluting waste</u> and it's <u>renewable</u> — the wind's never going to run out.

5) But there are <u>disadvantages</u>. You need about 5000 wind turbines to replace one coal-fired power station. Some people think that 'wind farms' spoil the view and the spinning blades cause noise pollution.

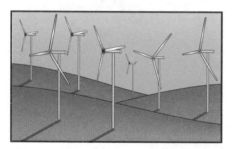

6) Another problem is that sometimes the wind isn't <u>strong enough</u> to generate any power. It's also impossible to increase supply when there's extra demand (e.g. when Coronation Street starts).

7) And although the wind is free, it's <u>expensive</u> to <u>set up</u> a wind farm, especially <u>out at sea</u>.

You can **capture** the Sun's energy using **solar cells**

1) <u>Solar cells</u> (<u>photocells</u>) generate <u>electricity directly</u> from sunlight. They generate <u>direct current</u> (DC) — the same as a <u>battery</u> (not like the <u>mains electricity</u> in your home, which is AC — alternating current).

2) Most solar cells are made of <u>silicon</u> — a <u>semiconductor</u>. When sunlight falls on the cell, the silicon atoms <u>absorb</u> some of the energy, knocking loose some <u>electrons</u>. These electrons then flow round a circuit.

3) The <u>power output</u> of a photocell depends on its <u>surface area</u> (the bigger the cell, the more electricity it produces) and the <u>intensity of the sunlight</u> hitting it (brighter light = more power). Makes sense.

The main problem is the visual impact

Wind turbines you love them or you hate them — I think they look really good and wouldn't mind seeing a few more. Others can't stand them, while some even have them put on the roof of their house.

Wind and Solar Energy

Solar cells — expensive to install, cheap to run

1) Solar cells are very <u>expensive initially</u>, but after that the energy is <u>free</u> and <u>running costs</u> are almost <u>nil</u>. And there's <u>no pollution</u> (although they use a fair bit of energy to manufacture in the first place).

2) Solar cells can only <u>generate</u> enough <u>electricity</u> to be useful if they have <u>enough sunlight</u> — which can be a problem at <u>night</u> (and in <u>winter</u> in some places). But the cells can be linked to <u>rechargeable batteries</u> to create a system that can <u>store energy</u> during the day for use at <u>night</u>.

3) Solar cells are often the best way to power <u>calculators</u> or <u>watches</u> that don't use much energy. They're also used in <u>remote places</u> where there's not much choice (e.g. deserts) and in satellites.

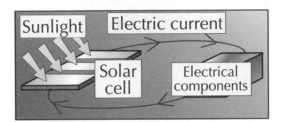

Passive solar heating — no complex mechanical stuff

Solar Panels

Solar panels are much less sophisticated than photocells — basically just <u>black water pipes</u> inside a <u>glass</u> box. The <u>glass</u> lets <u>heat</u> and <u>light</u> from the Sun in, which is then <u>absorbed</u> by the black pipes and heats up the water.

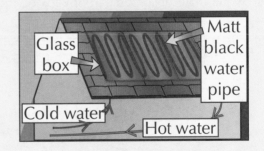

Cooking with Solar Power

If you get a <u>curved mirror</u>, then you can <u>focus</u> the Sun's light and heat. This is what happens in a solar oven.

All the radiation that lands on the curved mirror is focused right on your pan.

Unfortunately it's not too reliable here in Britain

And you can reduce the energy needed to <u>heat</u> a building if you build it sensibly in the first place — e.g. face the <u>windows</u> in a suitable direction. That's another example of passive solar heating.

Biomass

Biomass and wave energy sound pretty dull, but they're actually quite exciting possibilities.

Biomass is waste that can be burnt — plant and animal waste

1) Biomass is the general term for organic 'stuff' that can be burnt to produce electricity (e.g. farm waste, animal droppings, landfill rubbish, specially-grown forests...).

Plant material...

...rubbish...

...and waste.

2) The waste material is burnt in power stations to drive turbines and produce electricity. Or sometimes it's fermented to produce other fuels such as 'biogas' (mostly methane) or ethanol.

Biomass is carbon neutral...well, nearly

1) The plants that grew to produce the waste (or to feed the animals that produced the dung) would have absorbed carbon dioxide from the atmosphere as they were growing.

2) When the waste is burnt, this CO_2 is re-released into the atmosphere. So it has a neutral effect on atmospheric CO_2 levels.

Although this only really works if you keep growing plants at the same rate you're burning things, and if you ignore any fossil fuels used in transporting the fuel to the power station, etc.

Biomass schemes are cheap to set up

1) Set-up and fuel costs are generally low, since the fuel is usually waste, and the fuels can often be burnt in converted coal-fired power stations.

2) This process can make use of waste products, which could be great news for our already overflowing landfill sites. But the downside of using unsorted landfill rubbish, rather than just plant and animal waste, is that burning it can release nasty gases like sulphur dioxide and nitrogen oxide into the atmosphere.

Biomass — what a load of old rubbish

Biomass isn't the nicest of things — dead plants, rubbish and poo — mmm, my favourite. It could be a good way of sorting out our energy problems as there's lots of the stuff about already, so don't bin it.

Wave Energy

Wave power — *lots of little* **wave converters**

Don't confuse <u>wave power</u> with <u>tidal power</u> — they're <u>completely different</u>.

1) For wave power, you need lots of small <u>wave converters</u> located <u>around the coast</u>. As waves come in to the shore they provide an <u>up and down motion</u> which can be used to drive a <u>generator</u>.

2) There's <u>no pollution</u>. The main problems are <u>spoiling the view</u> and being a <u>hazard to boats</u>.

3) It's <u>fairly unreliable</u>, since waves tend to die out when the <u>wind drops</u>.

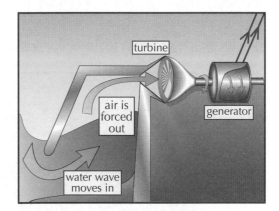

4) <u>Initial costs are high</u> but there are <u>no fuel costs</u> and <u>minimal running costs</u>. Wave power is unlikely to provide energy on a <u>large scale</u> but it can be <u>very useful</u> on <u>small islands</u>.

Tidal barrages — *using the* **Sun** *and* **Moon's gravity**

1) <u>Tidal barrages</u> are <u>big dams</u> built across <u>river estuaries</u>, with <u>turbines</u> in them. As the <u>tide comes in</u> it fills up the estuary to a height of <u>several metres</u>. This water can then be allowed out <u>through turbines</u> at a controlled speed. It also drives the turbines on the way in.

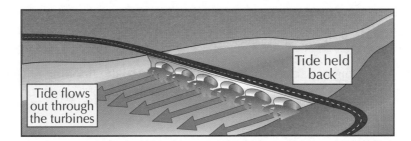

2) There's <u>no pollution</u>. The main problems are <u>preventing free access by boats</u>, <u>spoiling the view</u> and <u>altering the habitat</u> of the wildlife.

3) Tides are <u>pretty reliable</u>, but the <u>height</u> of the tide is <u>variable</u> so lower tides will provide <u>less energy</u> than higher ones.

4) <u>Initial costs are moderately high</u>, but there's <u>no fuel costs</u> and <u>minimal running costs</u>.

Don't get wave power and tidal power confused

I do hope you appreciate the <u>big big differences</u> between <u>tidal power</u> and <u>wave power</u>. They both involve salty sea water, sure — but there the similarities end. Smile and enjoy. And <u>learn</u>.

Hydroelectric Power

Here's another couple of renewables — learn the advantages and disadvantages. They both use water, but hydroelectricity generates electricity, whereas pumped storage only stores it. The clues are in the names — so don't mix them up.

Hydroelectricity uses water to produce electricity

1) <u>Hydroelectric power</u> often requires the <u>flooding</u> of a <u>valley</u> by building a <u>big dam</u>.

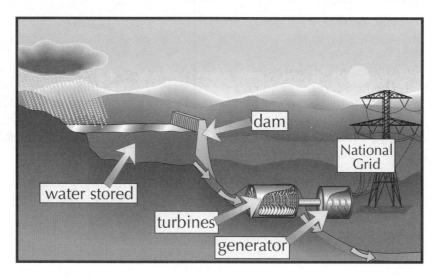

2) <u>Rainwater</u> is caught and allowed out <u>through turbines</u>.

3) The <u>movement</u> of the water turns the <u>turbine</u>, which turns a <u>generator</u> and produces electricity

4) There is <u>no pollution</u> from the running of a hydroelectric scheme.

Hydroelectricity impacts the environment

1) Hydroelectricity has a <u>big impact</u> on the <u>environment</u> due to the flooding of the valley (rotting vegetation releases methane and CO_2) and possible <u>loss of habitat</u> for some species (sometimes the loss of whole villages).

2) The reservoirs can also look very <u>unsightly</u> when they <u>dry up</u>. Location in <u>remote valleys</u> tends to avoid some of these problems.

3) A <u>big advantage</u> is <u>immediate response</u> to increased demand, and there's no problem with <u>reliability</u> except in times of <u>drought</u>.

4) <u>Initial costs are high</u>, but there's <u>no fuel</u> and <u>minimal running costs</u>.

Hydroelectric, unsightly and damaging to the environment, but clean

In Britain only a pretty <u>small percentage</u> of our electricity comes from <u>hydroelectric power</u> at the moment, but in some other parts of the world they rely much more heavily on it. For example, in the last few years, <u>99%</u> of <u>Norway's</u> energy came from hydroelectric power. 99% — that's huge!

Pumped Storage and Power Stations

Pumped storage gives extra supply just when it's needed

1) Most large power stations have huge boilers which have to be kept running all night even though demand is very low. This means there's a surplus of electricity at night.

2) It's surprisingly difficult to find a way of storing this spare energy for later use.

3) Pumped storage is one of the best solutions.

4) In pumped storage, 'spare' night-time electricity is used to pump water up to a higher reservoir.

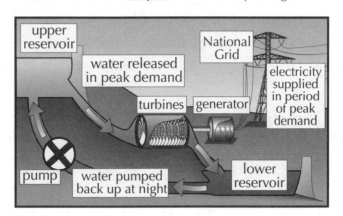

5) This can then be released quickly during periods of peak demand such as at teatime each evening, to supplement the steady delivery from the big power stations.

6) Remember, pumped storage uses the same idea as hydroelectric power but it isn't a way of generating power — but simply a way of storing energy which has already been generated.

Setting up a power station

Because coal and oil are running out fast, many old coal- and oil-fired power stations are being taken out of use. Mostly they're being replaced by gas-fired power stations. But gas is not the only option, as you really ought to know if you've been concentrating at all over the last few pages.

When looking at the options for a new power station, there are several factors to consider:

1) How much it costs to set up and run, how long it takes to build, how much power it can generate, etc.

2) Then there are also the trickier factors like damage to the environment and impact on local communities. And because these are often very contentious issues, getting permission to build certain types of power station can be a long-running process, and hence increase the overall set-up time.

Pumped storage — the clue's in the name, it **stores** energy

Electricity companies have one big problem — customers, they all want their electricity at the same time, so rude. Ok, so it's not the only problem — there's also the fact that coal and oil are running out, nobody wants a nuclear power station in their back yard and wind turbines are ugly. Great.

Warm-Up and Exam Questions

Warm-up questions first, then an exam question or two to practise.
Make the most of this page by working through everything carefully — it's all useful stuff.

Warm-up Questions

1) Give two ways in which using coal as an energy source causes problems.
2) Describe the major problems with using nuclear fuel to generate electricity.
3) Explain what is meant by biomass.
4) Describe the difference between a pumped storage scheme and a hydroelectric power scheme.
5) Give two reasons why solar cells are not widely used to generate electricity.

Exam Questions

1   Geothermal energy can be described as a renewable energy source.

(a)   What is geothermal energy?

(1 mark)

(b)   What does 'renewable energy' mean?

(1 mark)

(c)   Why is geothermal energy not used much in the U.K.?

(1 mark)

2   Which one of these energy sources does **not** depend on energy radiated by the Sun?

A   Wind

B   Wave

C   Biomass

D   Tidal

(1 mark)

3   The diagram shows a pumped storage plant. Match up the labels **1 – 4** on the diagram with the energy transfers **A**, **B**, **C** and **D**.

A   gravitational potential to kinetic

B   kinetic to electrical

C   electrical to kinetic

D   kinetic to gravitational potential

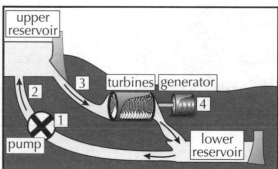

(4 marks)

Exam Questions

4 An old coal-fired power station has an output of 2 MW (2 million watts).
The electricity generating company plans to replace it with wind turbines which have
a maximum output of 4000 W each.

   (a) Calculate the minimum number of wind turbines required to replace the
        old power station.

        (1 mark)

   (b) Why might more wind turbines than this be needed in reality?

        (1 mark)

   (c) Suggest why some people might oppose the wind farm development.

        (2 marks)

5 Which is the most reliable method of generating electricity?

   A Tidal power

   B Wind power

   C Solar power

   D Wave power

        (1 mark)

6 The diagram shows a solar heating panel which is used to heat cold water in a house.

   (a) Describe how heat is transferred:

        (i)   from the Sun to the solar heating panel.

              (1 mark)

        (ii) from the hot water in the pipe to the colder
              water in the tank.

              (1 mark)

        (iii) throughout the water in the tank.

              (1 mark)

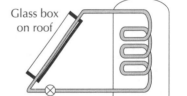

   (b) Explain why the pipes in the heating panel are painted black.

        (1 mark)

7 The inhabitants of a remote island do not have the resources or expertise to build a nuclear
power plant. They have no access to fossil fuels.

   (a) The islanders have considered using wind, solar and hydroelectric power to generate
        electricity. Suggest two other renewable energy resources they could use.

        (2 marks)

   (b) The islanders decide that both solar and hydroelectric power could reliably generate
        enough electricity for all their needs. Suggest two other factors they should consider
        when deciding which method of electricity generation to use.

        (2 marks)

Revision Summary for Section Seven

Phew... what a relief, you've made it to the end of yet another nice long section. This one's been fairly straightforward though — after all, about half of it just covered the pros and cons of different renewable energy resources. But don't kid yourself — there are definitely a shedload of facts to remember here and you need to know the lot of them. The best way to check that you know it all is to work your way through these revision questions — and make sure you go back and revise anything you get wrong.

1) Explain the difference between heat and temperature. What units are they each measured in?

2)* A rod of metal has a mass of 600 g. It's heated from 18 °C to 28 °C using 5400 J. Calculate the specific heat capacity of the metal.

3) Why does a graph showing the temperature of a substance as it's heated have two flat bits?

4)* The specific latent heat of water (for boiling) is 2.26×10^6 J/kg. How much energy does it take to boil dry a kettle containing 350 g of boiling water?

5) Describe the process that transfers heat energy through a metal rod. What is this process called?

6) Describe how the heat from the element is transferred throughout the water in a kettle. What is this process called?

7) Explain why solar hot water panels have a matt black surface.

8) The two designs of car engine shown are made from the same material. Which engine will transfer heat quicker? Explain why.

Engine A   Engine B

9) Name five ways of reducing the amount of heat lost from a house, and explain how they work.

10)*The following table gives some information about two different energy-saving light bulbs.

a) What is the payback time for light bulb A?
b) Which light bulb is more cost-effective over one year?

	Price of bulb	Annual saving
Light bulb A	£2.50	£1.25
Light bulb B	£3.00	£2.00

11) What does a thermogram show?

12) Name nine types of energy and give an example of each.

13) List the energy transformations that occur in a battery-powered toy car.

14) What is the useful type of energy delivered by a motor? In what form is energy wasted?

15) Write down the formula for calculating efficiency.

16)*What is the efficiency of a motor that converts 100 J of electrical energy into 70 J of useful kinetic energy?

17) What is meant by a non-renewable energy resource? Name four different non-renewable energy resources.

18) State two advantages and two disadvantages of using fossil fuels to generate electricity.

19) Outline two arguments for and two arguments against increasing the use, in the UK, of nuclear power.

20) Give two advantages and one disadvantage of using solar cells to generate electricity.

21) How do solar ovens focus the Sun's rays?

22) Describe how the following renewable resources are used to generate electricity. State one advantage and one disadvantage for each resource.

a) wind   b) biomass   c) geothermal energy
d) waves   e) the tide

*Answers on page 295.

Electric Current

Isn't electricity great — mind you it'll be a pain come exam time if you don't know the basics.

Electric current is a flow of electrons round a circuit

1) CURRENT is the flow of electrons round a circuit. (Electrons are negatively charged particles, see p98.)

2) VOLTAGE is the driving force that pushes the current round. Kind of like "electrical pressure".

3) RESISTANCE is anything in the circuit which slows the flow down.

4) There's a BALANCE: the voltage is trying to push the current round the circuit, and the resistance is opposing it — the relative sizes of the voltage and resistance decide how big the current will be:

> If you increase the VOLTAGE — then MORE CURRENT will flow.
>
> If you increase the RESISTANCE — then LESS CURRENT will flow.

It's just like the flow of water around a set of pipes

1) The current is simply like the flow of water.

2) Voltage is like the pressure provided by a pump which pushes the stuff round.

3) Resistance is any sort of constriction in the flow, which is what the pressure has to work against.

4) If you turn up the pump and provide more pressure (or "voltage"), the flow will increase.

5) If you put in more constrictions ("resistance"), the flow (current) will decrease.

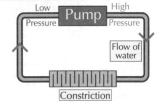

Electrons flow the opposite way to conventional current

We normally say current in a circuit flows from positive to negative. Alas, electrons were discovered long after that was decided and they turned out to be negatively charged. This means they actually flow from −ve to +ve, opposite to the flow of "conventional current".

AC keeps changing direction but DC does not

1) The mains electricity supply in your home is alternating current — AC. It keeps reversing its direction back and forth.

2) A cathode ray oscilloscope (CRO) can show current as a trace on a graph. The CRO trace for AC would be a wave. The voltage rises from zero to a peak positive value, then drops down to a 'peak' negative value, then back up to zero again, and so on. The frequency of the supply is how many of these waves you get per second.

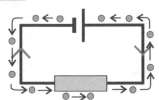

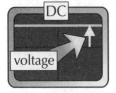

3) Direct current (DC) is different. It always flows in the same direction.

4) The CRO trace is a horizontal line. The voltage doesn't vary — so the current has a constant value too.

5) You get DC current from batteries and solar cells (see pages 212 & 213).

Current, Voltage and Resistance

Resistance, current and voltage are all closely linked. And if you don't believe me, you can easily check.

Investigating how current varies with voltage

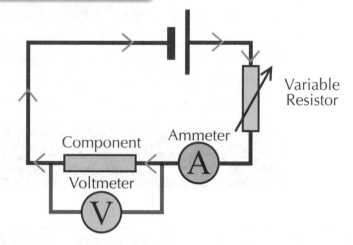

1) This circuit can be used to investigate how current varies with voltage for any component.

2) The ammeter is always connected in series with the component you're testing, to measure current flowing through it.

3) The voltmeter is always connected in parallel with the component to measure the voltage across it.

4) The supply voltage from the cell doesn't change. You use a variable resistor in the circuit, which you adjust to pick different values for the current, and for each value measure the voltage across the component.

You can use a V-I graph to look at resistance

Data from a circuit like the one above can be plotted it on a V-I (voltage-current) graph. This shows what happens to the current when you vary the voltage — as a bonus, the gradient shows the resistance.

Fixed-value resistor

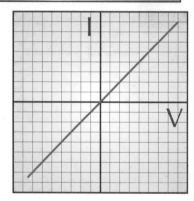

1) The resistance, R, of the component is constant (at a constant temperature).
2) If you plot voltage against current, you get a straight line — so you can see that the current is proportional to the voltage.

Filament Lamp

1) With a filament lamp, the resistance isn't constant — it increases as the current increases. This is because the bigger the current through the filament lamp, the hotter it gets. And as its temperature increases, its resistance increases.
2) The graph's a curve — the current is not proportional to the voltage.

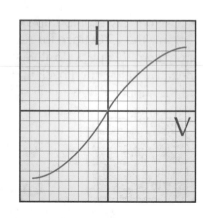

Current, Voltage and Resistance

You need to know the **resistance equation**

1) The <u>current</u> that flows through a component always depends on the <u>resistance</u> of the component and the <u>voltage</u> across it. There's a very simple equation for it:

$$\text{Voltage} = \text{Current} \times \text{Resistance}$$
(volts, V) (amps, A) (ohms, Ω)

or this much shorter version...

$$V = I \times R$$

> V is Voltage, and R is Resistance, but, oddly, I is for Current.

2) The equation above has <u>voltage</u> as its subject. But you might need to find <u>resistance</u> or <u>current</u> — which means rearranging the equation. You can do this with a <u>formula triangle</u>.

> Cover up the thing you're trying to find, then what you can still see is the formula you need to use.

*Conditions affect the resistance of **LDRs** and **thermistors***

1) A <u>light-dependent resistor</u> (LDR) has a resistance which depends on the <u>light level</u>.

2) The resistance is <u>highest</u> in total <u>darkness</u>. As the light gets <u>brighter</u>, the <u>resistance falls</u>. This makes it a useful device for various <u>electronic sensors</u>, e.g. <u>automatic night lights</u> and <u>burglar detectors</u>.

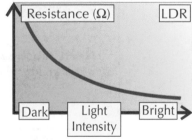

3) Thermistors have varying resistance which depends on the <u>temperature</u>.

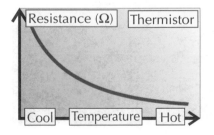

4) Most thermistors have a <u>high</u> resistance in <u>cold</u> conditions. As it gets <u>warmer</u>, the <u>resistance falls</u>. Thermistors are very useful as <u>temperature sensors</u> — e.g. in <u>car-engine</u> temperature gauges and <u>central-heating thermostats</u>.

In the end, you'll have to learn this — resistance is futile...

<u>V = I × R</u> is without doubt the most important equation in electronics. LEARN IT LEARN IT LEARN IT.

224

The Dynamo Effect

Generators use a pretty cool piece of physics to make electricity from the movement of a turbine. It's called electromagnetic (EM) induction — which basically means making electricity using a magnet.

> ELECTROMAGNETIC INDUCTION: The creation of a VOLTAGE (and maybe current) in a wire which is experiencing a CHANGE IN MAGNETIC FIELD.

The **dynamo effect** — move the **wire** or the **magnet**

1) Using electromagnetic induction to transform kinetic energy (energy of moving things) into electrical energy is called the dynamo effect. (In a power station, this kinetic energy is provided by the turbine.)

2) There are two different situations where you get EM induction:
   a)   An electrical conductor (a coil of wire is often used) moves through a magnetic field.

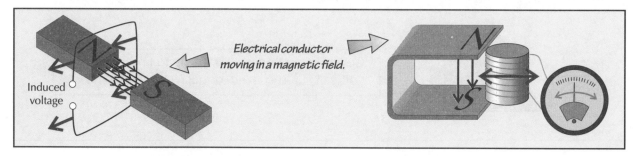

Electrical conductor moving in a magnetic field.

Induced voltage

   b)   The magnetic field through an electrical conductor changes (gets bigger or smaller or reverses).

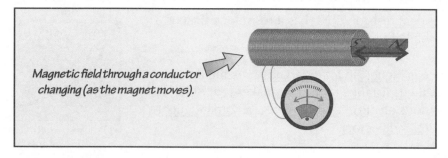

Magnetic field through a conductor changing (as the magnet moves).

3) If the direction of movement is reversed, then the voltage/current will be reversed too.

> To get a bigger voltage, you can increase...
> 1) The STRENGTH of the MAGNET
> 2) The number of TURNS on the COIL
> 3) The SPEED of movement

Learn the three ways of getting a bigger voltage

Who'd have thought you could make electricity from a magnet and a bit of wire — try it yourself if you don't believe it. While you're there try changing the magnet and the number of turns and things to see how the voltage changes. Once you get the hang of it, it'll stick in your head for the exam no trouble.

The Dynamo Effect

Generators move a coil in a magnetic field

1) Generators usually <u>rotate a coil</u> in a <u>magnetic field</u>.

2) Every half a turn, the current in the coil <u>swaps direction</u>.

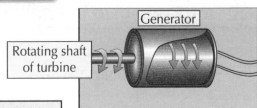

Generator

Rotating shaft of turbine

Think about one part of the coil... sometimes it's heading for the magnet's north pole, sometimes for the south — it changes every half a turn. This is what makes the current change direction.

3) This means that generators produce an <u>alternating (AC) current</u>. If you looked at the current (or voltage) on a display, you'd see something like this...

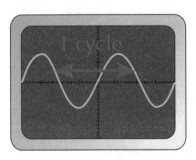

1 cycle

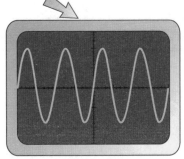

Turning the coil <u>faster</u> produces not only <u>more</u> peaks, but a <u>higher voltage</u> too.

4) The <u>frequency</u> of AC electrical supplies is the number of 'cycles' per second, and is measured in <u>hertz</u> (Hz). In the UK, electricity is supplied at 50 Hz (which means the coil in the generator at the power station is rotating 50 times every second).

5) Remember, this is completely different from the DC electricity supplied by batteries and solar cells. If you plotted that on a graph, you'd see something more like this...

6) <u>Dynamos</u> on bikes work slightly differently — they usually rotate the <u>magnet</u> near the coil. But the principle is <u>exactly the same</u> — they're still using EM induction.

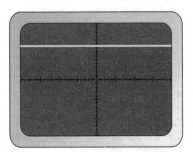

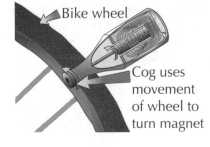

Bike wheel

Cog uses movement of wheel to turn magnet

Generators rotate a coil — dynamos rotate the magnet

EM induction sounds pretty hard, but it boils down to this — if a <u>magnetic field changes</u> (moves, grows, shrinks... whatever) somewhere near a <u>conductor</u>, you get <u>electricity</u>. It's a weird old thing, but important — this is how all our mains electricity is generated. We'd be in the dark without it.

Power Stations and the National Grid

Most of the electricity you use arrives via the national grid.

The **national grid** connects **power stations** to **consumers**

1) The national grid is the network of pylons and cables which covers the whole country.

2) It takes electricity from power stations to just where it's needed in homes and industry.

3) It enables power to be generated in a power station anywhere on the grid, and then supplied anywhere else on the grid.

All **power stations** are pretty much the **same...**

1) The aim of a power station is to convert one kind of energy (e.g. the energy stored in fossil fuels, or nuclear energy contained in the centre of atoms) into electricity.

2) Usually this is done in three stages...

① The first stage is to use the fuel (e.g. gas or nuclear fuel) to generate steam — this is the job of the boiler.

② The moving steam drives the blades of a turbine...

Nuclear reactors are just fancy boilers.

③ ...and this rotating movement from the turbine is converted to electricity by the generator (using electromagnetic induction — see the previous page).

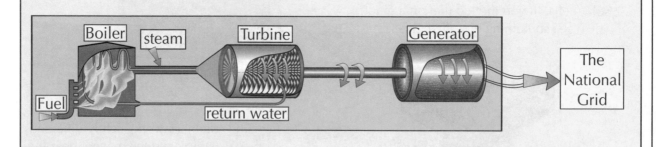

3) Most power stations are terribly inefficient — usually more than half the energy produced is wasted as heat and noise (though the efficiency of the power station depends a lot on the power source).

Most power stations are very very inefficient

Power stations might be big, but they're not all that clever — they boil water to make steam to turn a turbine. They also waste a lot of energy, which isn't great. So, invent a 100% efficient power station and you'll be rich and famous — I guess it's quite tricky though, or someone would've already done it.

Power Stations and the National Grid

The electricity generated by power stations is distributed around the country in the national grid. While it is being transmitted it has a higher voltage than in the power station and in your house.

Electricity is *transformed* to *high voltage* before *distribution*

1) To transmit a lot of electrical power, you either need a <u>high voltage</u> or a <u>high current</u> (see page 230 for more info about why). But... a higher current means your cables get hot, which is very inefficient (all that heat just goes to waste).

2) It's much cheaper to <u>increase</u> the <u>voltage</u>. So before the electricity is sent round the country, the voltage is <u>transformed</u> (using a <u>transformer</u>) to <u>400 000 V</u>. This keeps the current very <u>low</u>, meaning less wasted energy.

3) To increase the voltage, you need a <u>step-up transformer</u>.

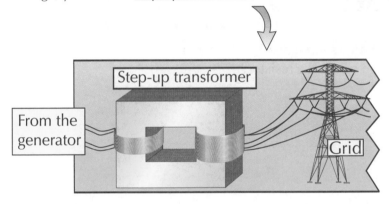

4) Even though you need big <u>pylons</u> with <u>huge</u> insulators (as well as the transformers themselves), using a high voltage is the <u>cheapest</u> way to transmit electricity.

5) To bring the voltage down to <u>safe usable levels</u> for homes, there are local <u>step-down transformers</u> scattered round towns — for example, look for a little fenced-off shed with signs all over it saying "Keep Out" and "Danger of Death".

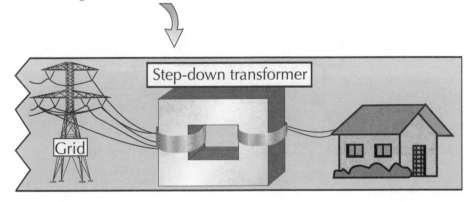

6) This is the main reason why mains electricity is AC — transformers <u>only work</u> on <u>AC</u>.

Fuel, boiler, steam, turbine, generator, transformer, grid, toaster

If you had your own solar panel or wind generator, you could sell back any surplus electricity to the national grid. So if you don't use much electricity, but you generate a lot of it, you can actually make money instead of spending it. Nice trick if you can do it. Shame solar panels cost a fortune...

Warm-Up and Exam Questions

This stuff isn't everyone's cup of tea. But once you get the knack of it, through lots of practice, you'll find the questions aren't too bad. Which is nice.

Warm-up Questions

1) What is electrical current?
2) Explain the difference between AC and DC.
3) What happens to the resistance of a thermistor as it gets cooler?
4) What is electromagnetic induction?
5) What is the National Grid?
6) Write down the equation that relates resistance, current and voltage.

Exam Questions

1   A current of 2.5 A flows through a 10 W resistor.
    What is the voltage across the resistor?

    A   25 V

    B   10 V

    C   4 V

    D   2.5 V

    (1 mark)

2   The circuit shown can be used to find the resistance of a component.

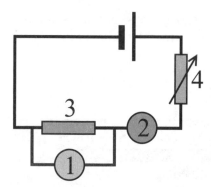

    Match the descriptions **A**, **B**, **C** and **D** to the parts **1-4** on the diagram.

    A   Variable resistor

    B   Voltmeter

    C   Ammeter

    D   Component to be tested

    (4 marks)

Exam Questions

3   Electricity is transmitted over long distances in the National Grid at

    A   low voltage and high current.

    B   low voltage and low current.

    C   high voltage and high current.

    D   high voltage and low current.

(1 mark)

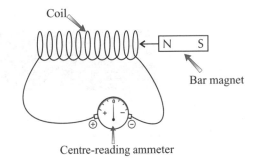

Coil

Bar magnet

Centre-reading ammeter

4   The diagram shows a coil of wire connected to an ammeter. Tim moves a bar magnet into the coil as shown. The pointer on the ammeter moves to the left.

    (a) Explain why the pointer moves.

(1 mark)

    (b) What could Tim do to get the ammeter's pointer to move to the right?

(1 mark)

    (c) How could Tim get a larger reading on the ammeter?

(1 mark)

    (d) What reading will the ammeter show if Tim holds the magnet still inside the coil?

(1 mark)

5   The diagram shows how heat energy from burning coal is used to generate electricity.

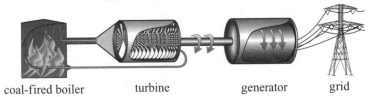

coal-fired boiler   turbine   generator   grid

    (a) In what form is the energy in coal stored?

(1 mark)

    (b) Coal has an energy content of approximately 25 000 kJ/kg.

        (i) How much heat energy is released in burning 1000 kg of coal?

(1 mark)

        (ii) Calculate the percentage efficiency of the power station if 10 000 MJ of electricity is generated using 1000 kg of coal.

(1 mark)

Electrical Power

This page is about the <u>power</u> of electrical appliances.

Electrical power is the rate of transfer of electrical energy

1) Electrical appliances are useful because they take in <u>electrical energy</u> and <u>convert it</u> into <u>other forms of energy</u>, e.g. a light bulb turns <u>electrical</u> energy into <u>light</u> and <u>heat</u> energy.

2) Converting energy from one form to another is called <u>transfer of energy</u>.

3) The electrical <u>power</u> of an appliance tells you how <u>quickly</u> it transfers electrical energy. The <u>units</u> of power are watts (W) or kilowatts (kW). 1 kilowatt = 1000 watts.

> ### <u>ELECTRICAL POWER</u> is the <u>Rate of Transfer</u> of <u>Electrical Energy</u>.

4) The <u>higher</u> the power of your appliance, the <u>more energy</u> is transferred every second. So a 3 kW kettle boils water <u>faster</u> than a 2 kW kettle, and a 100 W light bulb is <u>brighter</u> than a 60 W bulb.

Power = current × voltage

There's a nice easy equation for the <u>power</u> of an appliance...

> ### POWER = CURRENT × VOLTAGE
> (watts, W)   (amps, A)   (volts, V)

...or this much shorter version. $P = I \times V$

As usual, you need to practise <u>rearranging</u> the equation too.

$$\frac{P}{I \times V}$$

EXAMPLE: Anna's hairdrier has a power rating of 1.1 kW. She plugs the hairdrier into the 230 V mains supply. What is the current through the hairdrier?

You'll need to change the <u>units</u> — in this formula, power has to be in <u>watts</u>, not kilowatts.

ANSWER: You're trying to find <u>current</u>, so you need to rearrange the equation. Using the formula triangle, I = P ÷ V = 1100 ÷ 230 = <u>4.8 A</u>.

Electrical Power

All the electricity you use has to be paid for. Your electricity meter counts how many units of electricity you use and the electricity company multiplies this by the cost of each one to work out your bill.

Kilowatt-hours (kWh) are "UNITS" of energy

Your electricity meter records how much energy you use in units of kilowatt-hours, or kWh.

> A KILOWATT-HOUR is the amount of electrical energy converted by a 1 kW appliance left on for 1 HOUR.

Appliances with high power ratings cost more to run

The higher the power rating of an appliance, and the longer you leave it on, the more energy it consumes, and the more it costs. Learn (and practise rearranging) this equation too...

> UNITS OF ENERGY = POWER × TIME
> (in kWh)   (in kW) (in hours)

Write out a formula triangle if it helps.

And this one (but this one's easy):

> COST = NUMBER OF UNITS × PRICE PER UNIT

EXAMPLE: Find the cost of leaving a 60 W light bulb on for 30 minutes if one kWh costs 10p.

ANSWER: Energy (in kWh) = Power (in kW) × Time (in hours)
= 0.06 kW × ½ hr = 0.03 kWh
Cost = number of units × price per unit = 0.03 × 10p = 0.3p

Watt is the unit of power?
Get a bit of practice with the equations in those lovely burgundy boxes, and try these questions:
1) A kettle draws a current of 12 A from the 230 V mains supply. Calculate its power rating.
2) With 0.5 kWh of energy, for how long could you run the kettle?   Answers on p295.

Electrical Safety Devices

Electricity is dangerous. Just watch out for it, that's all.

Electrical cables usually have **live**, **neutral** and **earth** wires

In most electrical appliances, like toasters, say, the electrical cable has <u>three copper wires</u> inside it. Each wire is covered with an <u>insulating sheath</u> — which is a <u>plastic</u>, in a different colour for each wire.

The <u>brown</u> one is the <u>LIVE WIRE</u>. With a normal AC supply, this wire alternates between a <u>HIGH +VE AND –VE VOLTAGE</u>. The live wire has a <u>fuse</u> (or a trip switch) in it.

The <u>blue</u> one is the <u>NEUTRAL WIRE</u>. This wire is always at <u>0 V</u>. Current normally flows in and out of the appliance through the <u>live</u> and <u>neutral</u> wires.

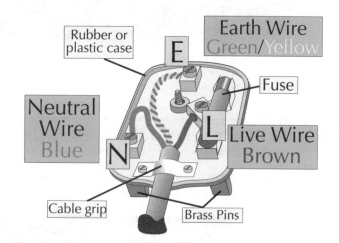

Rubber or plastic case

Earth Wire Green/Yellow

Fuse

Neutral Wire Blue

Live Wire Brown

Cable grip

Brass Pins

The <u>green and yellow</u> wire is the <u>EARTH WIRE</u>. One end of this wire is connected to the <u>earth</u> — it's usually clipped on to your cold water pipe, which comes from underground. The other end of the wire is connected to the <u>metal casing</u> of your appliance — so the casing is 'earthed'.

Residual Current Circuit Breakers help **prevent electrocution**

Sometimes, you can protect yourself with a <u>Residual Current Circuit Breaker</u> (<u>RCCB</u>) instead of a fuse and an earth wire. RCCBs work slightly differently.

1) Normally exactly the <u>same current</u> flows through the <u>live</u> and <u>neutral</u> wires.

2) However, if somebody <u>touches</u> the live wire, a current will flow <u>through them</u> to the <u>earth</u>. So now the neutral wire is carrying <u>less current</u> than the live wire.

3) The RCCB detects this <u>difference</u> in current and <u>quickly cuts off the power</u>.

Fuses — you'll find them in exams and kettles

It's really important to learn the colour of each of the wires — not just for the exam but for real life too. I like to remember them like this... green is for grass, and where does grass grow — in the earth of course. Now blue and neutral both have e and u in so they must go together, which just leaves the live wire — if it's alive it must be a worm, so is brown. Easy peasy... ish.

Electrical Safety Devices

Earth wires and fuses can also protect you from electric shocks

The earth wire doesn't normally have any current flowing through it.
The earth wire and fuse are just there for safety — and they work together like this:

1) If your toaster develops a fault, the live wire could touch the metal casing of the toaster.
   The outside of the toaster would then be at high voltage, and potentially dangerous,
   because you could easily touch it.

2) BUT the metal case is connected to the earth (which is at 0 V) by the earth wire — which
   is just a normal copper wire with low resistance. So a very big current flows in through
   the live wire, through the metal casing and out through the earth wire.

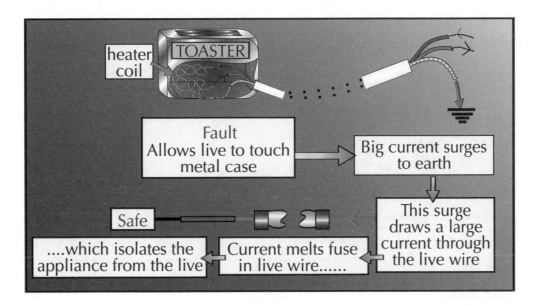

heater coil — TOASTER

Fault
Allows live to touch
metal case → Big current surges
to earth

This surge
draws a large
current through
the live wire

Safe ← ....which isolates the
appliance from the live ← Current melts fuse
in live wire......

3) This large current quickly melts the fuse in the live wire (or flips the trip switch)
   — and cuts off the high voltage supply.

4) This isolates the whole appliance from the high voltage supply, making it impossible to get
   an electric shock from the case of the appliance. It also prevents the risk of fire caused by
   the heating effect of a large current.

> Fuses have to be replaced once they've melted, whereas
> RCCBs can easily be reset by flicking a switch on the device.
> This makes RCCBs more convenient than fuses.

Loads of details to learn I'm afraid

Not all electrical appliances have to be earthed. If the appliance has a plastic casing and no metal
parts exposed, it's said to be double insulated. Anything with double insulation just needs a live and a
neutral wire. Household products like hairdriers and vacuum cleaners usually have double insulation.

234

Warm-Up and Exam Questions

I know that you'll be champing at the bit to get into the exam questions,
but these warm-up questions are invaluable for getting the basics straight first.

Warm-up Questions

1) What is electrical power?
2) What unit is used by electricity suppliers to charge customers for electricity usage?
3) Explain why electrical appliances with metal cases should be earthed.
4) Explain how a fuse works.
5) Why are RCCBs more convenient than fuses?

Exam Questions

1   A 1200 W toaster is used with the 230 V mains electricity supply.
    Calculate the current it draws.

    A   0.19 A

    B   1 A

    C   5.22 A

    D   13 A

    (1 mark)

2   There are three wires in the plug connected to Tom's washing machine.
    The wire with brown insulation is the

    A   earth wire

    B   live wire

    C   neutral wire

    D   negative wire

    (1 mark)

3   Sarah moves into her new flat and wants to estimate her electricity bill.
    She writes down the power of all her electrical appliances and estimates the time
    they're on for each day. She finds that she uses an average of 8.75 kWh per day.

    (a) How much energy, in joules, does she use per day?

    (1 mark)

    (b) If electricity costs 14.4p per kWh, what will her daily cost be?

    (1 mark)

    (c) The electricity company sends Sarah a bill every three months.
         Calculate her likely bill for 90 days' electricity supply.

    (1 mark)

    (d) A new electricity supplier charges 12p per kWh for electricity.
         How much would Sarah save per day if she switched to this supplier?

    (1 mark)

SECTION EIGHT — ELECTRICITY AND WAVES

Waves — The Basics

Waves transfer <u>energy</u> from one place to another without transferring any <u>matter</u> (stuff).

Waves have amplitude, wavelength and frequency

1) The <u>amplitude</u> is the displacement from the <u>rest position</u> to the <u>crest</u> (NOT from a trough to a crest).

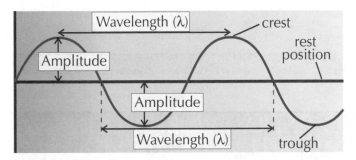

2) The <u>wavelength</u> is the length of a <u>full cycle</u> of the wave, e.g. from <u>crest to crest</u>.

3) <u>Frequency</u> is the <u>number of complete waves</u> passing a certain point <u>per second</u>. Frequency is measured in hertz (Hz). 1 Hz is <u>1 wave per second</u>.

Transverse waves have sideways vibrations

<u>Most waves</u> are <u>transverse</u>:

1) <u>Light</u> and <u>all other EM waves</u>.

2) <u>Ripples</u> on water.

3) <u>Waves</u> on <u>strings</u>.

4) A <u>slinky spring</u> wiggled up and down.

In <u>TRANSVERSE waves</u> the vibrations are at <u>90°</u> to the <u>DIRECTION OF TRAVEL</u> of the wave.

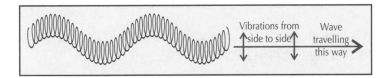

Longitudinal waves have vibrations along the same line

Examples of <u>longitudinal waves</u> are:

1) <u>Sound waves</u> and <u>ultrasound</u>.

2) <u>Shock waves</u>, e.g. seismic waves (see p253).

3) A <u>slinky spring</u> when you <u>push</u> the end.

In <u>LONGITUDINAL</u> waves the vibrations are along the <u>SAME DIRECTION</u> as the wave is travelling.

Oscilloscopes even show longitudinal waves as transverse — just so you can see what's going on.

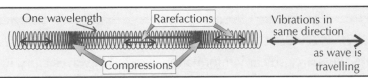

Waves — The Basics

These types of wave come up loads in physics so you'd better get them sorted in your head.

There are **seven types** of electromagnetic (EM) waves

Electromagnetic (EM) radiation is all around you. There are seven basic types of electromagnetic waves:

Wavelength

RADIO WAVES	MICRO WAVES	INFRA RED	VISIBLE LIGHT	ULTRA VIOLET	X-RAYS	GAMMA RAYS
$1m\text{-}10^4m$	$10^{-2}m$ (3cm)	$10^{-5}m$ (0.01mm)	$10^{-7}m$	$10^{-8}m$	$10^{-10}m$	$10^{-12}m$

Frequency

Electromagnetic waves travel at the **same speed**

1) All forms of electromagnetic radiation travel at the same speed through a vacuum.

2) Waves with a shorter wavelength have a higher frequency (see next page for why).

3) As a rule the EM waves at each end of the spectrum tend to be able to pass through material, while those nearer the middle are absorbed.

> When EM radiation is absorbed, it can cause:
>
>   i) heating,
>
>   ii) a tiny AC current with the same frequency as the radiation.

4) Also, the ones with higher frequency (shorter wavelength), like X-rays, tend to be more dangerous to living cells. That's because they have more energy. See page 240 for more information.

5) About half the EM radiation we receive from the Sun is visible light. Most of the rest is infrared (heat), with some UV thrown in. UV is what gives us a suntan (see page 241).

Remember: all waves carry energy without transferring matter

Waves carry energy, but can also carry information — e.g. EM waves carry TV signals, sound waves carry speech, and water waves carry... um... boats. Anyway, get learning about the properties of electromagnetic waves. Quite a straightforward page, so make the most of it.

Wave Behaviour

This stuff is true for __all__ waves — not just EM ones...

Wave speed = frequency × wavelength

You need to learn this equation (it's not given in the exam) and <u>practise using it</u>.

$$\text{Speed} = \text{Frequency} \times \text{Wavelength}$$
$$\text{(m/s)} \quad\quad \text{(Hz)} \quad\quad\quad\quad \text{(m)}$$

Or you can use the shortened version:

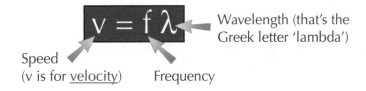

$$v = f\,\lambda$$

Wavelength (that's the Greek letter 'lambda')

Speed (v is for <u>velocity</u>)   Frequency

<u>EXAMPLE:</u> A radio wave has a frequency of 92.2×10^6 Hz.
Find its wavelength.
(The speed of all EM waves is 3×10^8 m/s.)

<u>ANSWER:</u> You're trying to find λ using f and v, so you've got to rearrange the equation.
So $\lambda = v \div f = 3 \times 10^8 \div 9.22 \times 10^7 = \underline{3.25 \text{ m}}$.

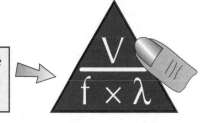

All EM waves travel at the same speed in a vacuum — so <u>waves with a high frequency must have a short wavelength</u>.

Waves can interfere with each other

1) When two or more waves of a <u>similar frequency</u> come into contact, they can create one combined signal with a new <u>amplitude</u>.

2) This is called <u>interference</u>. You get it when two radio stations transmit on similar frequencies.

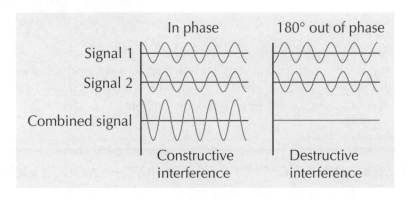

Signal 1
Signal 2
Combined signal

In phase
180° out of phase

Constructive interference
Destructive interference

Wave Behaviour

Waves travel along by themselves quite happily — but what happens when they meet an obstacle...

All waves can be reflected, refracted and diffracted

When waves arrive at an obstacle (or meet a new material), their direction of travel can be changed.

1) The waves might be <u>reflected</u> — so the waves 'rebound off' the material.

2) They could be <u>refracted</u> — which means they go through the new material but <u>change direction</u> (page 243).

3) Or they could be <u>diffracted</u> — this means the waves 'bend round' obstacles, causing the waves to spread out. This allows waves to 'travel round corners'.

Here's how diffraction works:

- All waves (<u>diffract</u>) <u>spread out</u> at the edges when they pass through a <u>gap</u> or <u>past an object</u>.

- The amount of diffraction depends on the size of the gap relative to the wavelength of the wave. The <u>narrower the gap</u>, or the <u>longer the wavelength</u>, the <u>more</u> the wave spreads out.

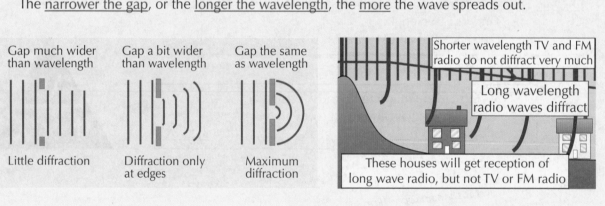

Gap much wider than wavelength — Little diffraction

Gap a bit wider than wavelength — Diffraction only at edges

Gap the same as wavelength — Maximum diffraction

Shorter wavelength TV and FM radio do not diffract very much

Long wavelength radio waves diffract

These houses will get reception of long wave radio, but not TV or FM radio

4) As well as changing a wave's direction of travel, an obstacle can have an effect on the wave itself. The obstacle might:

    i)   <u>absorb</u> the wave,

    ii) <u>transmit</u> it (let it pass through),

    iii) or <u>reflect</u> it.

Any <u>combination</u> of these three is also possible (so you can look at stuff in shop windows <u>and</u> check your hair). What actually happens depends on the <u>wavelength</u> of the radiation, the <u>material</u> the obstacle is made of, and what the <u>surface</u> of this material is like (its colour, shininess, etc. — see page 201 for more info).

Reflection, refraction, diffraction — you need to know them all

In 1588, <u>beacons</u> were used on the south coast of England to <u>relay</u> the information that the Spanish Armada was approaching. As we know, <u>light</u> travels as <u>electromagnetic waves</u>, so this is an early example of transferring information using electromagnetic radiation — or <u>wireless communication</u>.

Warm-Up and Exam Questions

Warm-up Questions

1) What is meant by the frequency of a wave?
2) Explain how waves can 'interfere' with each other, and how this affects the signals they're carrying.
3) Waves spread out when they travel through gaps. What is this effect called?
4) Give an example of a longitudinal wave.

Exam Questions

1   Which of the following types of electromagnetic wave has the highest frequency?

   A   Radio waves

   B   Ultraviolet

   C   Gamma rays

   D   Microwaves

   (1 mark)

2   The speed of electromagnetic waves in a vacuum is approximately 300 000 000 m/s. Use this figure to calculate the wavelength of a 100 MHz radio signal.

   A   3 m

   B   300 m

   C   30 000 m

   D   3 000 000 m

   (1 mark)

3   Look at this displacement-time graph for a water wave.

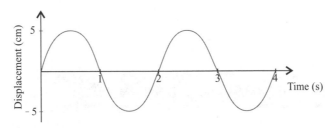

   (a)   What is the amplitude of this wave?

   (1 mark)

   (b)   Calculate the frequency of the wave.

   (1 mark)

   (c)   If the frequency of the wave doubles but its speed stays the same, what will happen to its wavelength?

   (1 mark)

Dangers of EM Radiation

Electromagnetic radiation can be dangerous. (So it'll probably be banned one of these days... sigh.)

Some radiations are **more harmful** than others

When EM radiation enters <u>living tissue</u> — like <u>you</u> — it's often harmless, but sometimes it creates havoc.

1) Some EM radiation mostly <u>passes through soft tissue</u> without being absorbed — e.g. radio waves.

2) Other types of radiation are absorbed and cause <u>heating</u> of the cells — e.g. microwaves.

3) Some radiations cause <u>cancerous changes</u> in living cells — e.g. UV can cause skin cancer.

4) Some types of EM radiation can actually <u>destroy cells</u> — as in 'radiation sickness' after a nuclear accident.

Higher frequency EM radiation is usually **more dangerous**

1) As far as we know, <u>radio</u> waves are pretty harmless.

2) This is because the <u>energy</u> of any electromagnetic wave is <u>directly proportional</u> to its <u>frequency</u>.

3) <u>Visible light</u> isn't harmful unless it's really <u>bright</u>. People who work with powerful <u>lasers</u> (very intense light beams) need to wear eye protection.

4) <u>Infrared</u> can cause <u>burns</u> or <u>heatstroke</u> (when the body overheats) — but they're <u>easily avoidable</u> risks.

> In general, waves with <u>lower frequencies</u> (like radio) are <u>less harmful</u> than <u>high frequency</u> waves like X-rays and gamma rays.

> <u>Higher frequency</u> waves have <u>more energy</u>. And it's the <u>energy</u> of a wave that does the <u>damage</u>.

Microwaves — **may** or **may not** be **harmful**

1) Some wavelengths of microwaves are <u>absorbed</u> by <u>water</u> molecules and <u>heat</u> them up. If the water in question happens to be in <u>your cells</u>, you might start to <u>cook</u>.

2) Mobile phone networks use microwaves, and some people think that using your mobile a lot, or living near a <u>mast</u>, could damage your <u>health</u>. There isn't any conclusive proof either way yet.

In general, the more energy a wave has, the more damage it can do

If you can remember the electromagnetic spectrum, this page should be a doddle. Just think about the two ends of the spectrum — would you rather have some tuneful radio waves or scary gamma rays? I'd go for the radio waves any day — the dangers of the other types of radiation fit nicely in between.

Dangers of EM Radiation

Information about the dangers of too much Sun is everywhere nowadays, but getting a tan can't be that bad for you... can it?

Ultraviolet radiation can cause *skin cancer*

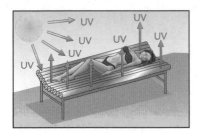

1) If you spend a lot of time in the sun, you'll get a tan and maybe sunburn (with attractive peeling).

2) The more time you spend in the sun, the more chance you also have of getting skin cancer. This is because the Sun's rays include ultraviolet radiation (UV) which damages the DNA in your cells.

3) Dark skin gives some protection against UV rays — it absorbs more UV radiation, stopping it from reaching the more vulnerable tissues deeper in the body.

4) Everyone should protect themselves from overexposure to the Sun, but if you're pale skinned, you need to take extra care, and use a sunscreen with a higher Sun Protection Factor (SPF).

> An SPF of 15 means you can spend 15 times as long in the sun as you otherwise could without burning (if you keep reapplying the sunscreen).

5) The gas inside fluorescent tubes (often used for kitchen and office lighting) emits UV radiation. Special coatings are used on lamps to absorb the UV and emit visible light instead.

The *ozone layer* protects us from *UV radiation*

The ozone layer absorbs some UV rays from the Sun — so it reduces the amount of UV radiation reaching the Earth's surface. Recently, the ozone layer has got thinner because of pollution from CFCs.

- CFCs are gases which react with ozone molecules and break them up. This depletion of the ozone layer allows more UV rays to reach us at the surface of the Earth.

- We used to use CFCs all the time — e.g. in hairsprays and in the coolant for fridges — they're now banned or restricted because of their environmental impact.

Staying in the sun for too long can cause cancer

There's no point being paranoid — a little bit of sunshine won't kill you (in fact it might do you good). But don't be daft... getting cancer from sunbathing for hours on end is just stupid. As for CFCs — that's chlorofluorocarbons by the way — they're mostly banned now but the ozone layer is still being damaged.

Warm-Up and Exam Questions

You must be getting used to the routine by now — the warm-up questions get you, well, warmed up, and the exam questions give you some idea of what you'll have to cope with on the day.

Warm-up Questions

1) Explain how high frequency electromagnetic radiation can damage cells.
2) Why can it be dangerous for living cells to absorb infrared radiation?
3) Why is the use of CFCs now restricted?
4) Which type of radiation does the ozone layer absorb?

Exam Questions

1   The radiation used in mobile phone networks

   A   is definitely safe

   B   has a higher frequency than visible light

   C   may cause damage to cells

   D   is known to cause health problems

   (1 mark)

2   Match the words **A**, **B**, **C** and **D** with the spaces **1-4** in the sentences.

   A   Visible light

   B   Radio

   C   Infrared

   D   Ultraviolet

   Using sunscreen will give you some protection from ...**1**... radiation emitted by the Sun.

   Grills use ...**2**... radiation to cook food.

   ...**3**... is not usually dangerous unless it's very bright e.g. lasers.

   ...**4**... waves are not known to have any ill effects on the body.

   (4 marks)

3   (a) Give one hazard and one practical use of

       (i)   microwave radiation

   (2 marks)

       (ii) gamma radiation

   (2 marks)

   (b) Give one way in which you can reduce your exposure to UV radiation from the Sun.

   (1 mark)

Refraction

All waves can be <u>refracted</u> — it's a fancy way of saying '<u>made to change direction</u>'.

Waves can be **refracted**

1) Waves travel at <u>different speeds</u> in substances which have <u>different densities</u>. EM waves travel more <u>slowly</u> in <u>denser</u> media (usually). Sound waves travel faster in <u>denser</u> substances.

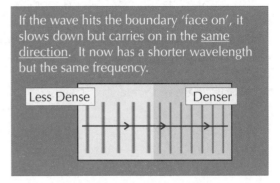

If the wave hits the boundary 'face on', it slows down but carries on in the <u>same</u> <u>direction</u>. It now has a shorter wavelength but the same frequency.

Less Dense — Denser

2) So when a wave crosses a boundary between two substances (from glass to air, say) it <u>changes speed</u>.

3) When light shines on a glass <u>window pane</u>, some of the light is reflected, but a lot of it passes through the glass and gets <u>refracted</u> as it does so.

4) As the light passes from the air into the glass (a <u>denser</u> medium), it <u>slows down</u>. This causes the light ray to bend <u>towards</u> the normal.

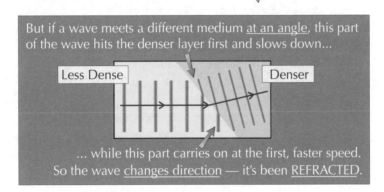

But if a wave meets a different medium <u>at an angle</u>, this part of the wave hits the denser layer first and slows down...

Less Dense — Denser

... while this part carries on at the first, faster speed. So the wave <u>changes direction</u> — it's been <u>REFRACTED</u>.

5) When the light reaches the 'glass to air' boundary on the other side of the window, it <u>speeds up</u> and bends <u>away</u> from the normal. (Some of the light is also <u>reflected</u> at each of the boundaries.)

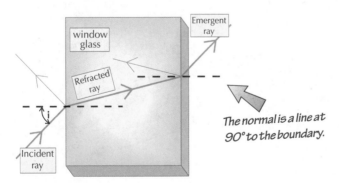

Emergent ray

window glass

Refracted ray

i

Incident ray

The normal is a line at 90° to the boundary.

Make sure you can describe refraction and all the diagrams

Refraction sounds a bit like reflection — but they're not the same thing, so try not to mix them up. Refraction is all about waves travelling through different things, whereas reflection is where the waves bounce off a surface and come right back at you. See, not the same thing at all.

Refraction

Total internal reflection happens above the critical angle

1) <u>Total internal reflection</u> can only happen when a wave travels <u>through a dense substance</u> like glass or water or perspex <u>towards a less dense</u> substance like air.

2) It all depends on whether the angle of incidence (i.e. the angle it hits at) is <u>bigger</u> than the <u>critical angle</u>...

If the angle of incidence (i) is:

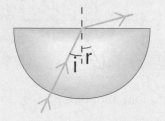

The <u>angle of incidence</u> (i) and the <u>angle of reflection</u> (r) are always measured from the <u>normal</u> (a line at right angles to the surface).

...<u>LESS</u> than Critical Angle:-
Most of the light <u>passes out</u> but a <u>little</u> bit of it is <u>internally reflected</u>.

Angle of reflection, r, equals the <u>angle of incidence</u>, i.

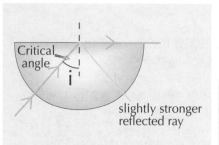

...<u>EQUAL</u> to Critical Angle:-
The emerging ray comes out <u>along the surface</u>. There's quite a bit of <u>internal reflection</u>.

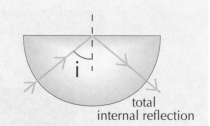

total internal reflection

...<u>GREATER</u> than Critical Angle:-
<u>No light comes out</u>. It's <u>all</u> internally reflected,
i.e. <u>total internal reflection</u>.

3) Different materials have different critical angles. The critical angle for <u>glass</u> is about 42°.

4) <u>Optical fibres</u> work by bouncing waves off the sides of a thin <u>inner core</u> of glass or plastic using total internal reflection. The wave enters one end of the fibre and is reflected repeatedly until it emerges at the other end.

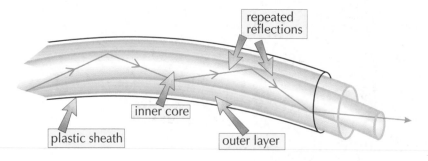

repeated reflections

inner core

plastic sheath

outer layer

5) Optical fibres <u>can be bent</u>, but <u>not sharply</u>, or the <u>angle of incidence</u> might fall <u>below</u> the critical angle.

Total internal reflection can be amazingly useful

Optical fibres are a good way to send data over long distances — the EM waves travel fast, and they can't be tapped into or suffer interference (unlike a signal that's <u>broadcast</u> from a transmitter, like radio).

Radio Waves

Radio waves have <u>long wavelengths</u> and <u>low frequencies</u> — making them great for communications.

Radio waves *are used mainly for* communications

1) <u>Radio waves</u> (EM radiation with wavelengths longer than about 10 cm) and some <u>microwaves</u> are good at transferring information <u>long distances</u>.

2) This is partly because they don't get <u>absorbed</u> by the Earth's atmosphere as much as waves in the <u>middle</u> of the EM spectrum (like heat, for example), or those at the high-frequency end of the spectrum (e.g. gamma rays or X-rays — though these would be too dangerous to use anyway).

3) But the different wavelengths have different properties, and are used in different ways...

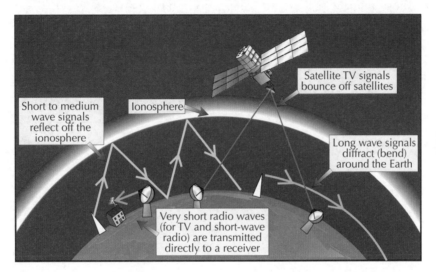

- <u>Long-wave radio</u> (wavelengths of <u>1 – 10 km</u>) can be transmitted from London, say, and received halfway round the world, because long wavelengths <u>diffract</u> around the curved surface of the Earth.

- The radio waves used for <u>TV and FM radio</u> transmissions, and the microwaves used for mobile phone communications, have very short wavelengths (10 cm – 10 m). These signals don't diffract much — to get reception, you must be in <u>direct sight of the transmitter</u>. This is why mobile phone transmitters are positioned on <u>hill tops</u> and fairly <u>close to one another</u>.

  Signals for satellite TV go through the atmosphere and 'bounce' off satellites:

  > 1) A <u>transmitter</u> on Earth sends the signal up into space...
  >
  > 2) ...where it's picked up by the <u>satellite receiver dish</u> orbiting thousands of kilometres above the Earth. The satellite transmits the signal <u>back to Earth</u>...
  >
  > 3) ...where it's picked up by a receiving <u>satellite dish</u>.

- <u>Short-wave radio</u> signals (wavelengths of about <u>10 m – 100 m</u>) can, like long wave, be received at <u>long distances</u> from the transmitter. That's because they're <u>reflected</u> from the <u>ionosphere</u> — an <u>electrically charged layer</u> in the Earth's upper atmosphere.

That's why hilly areas can get French radio better than Radio 4

Radio waves can be long, medium or short wave — but don't be fooled, short-wave radio waves still have a longer wavelength than other sorts of electromagnetic waves. There's short and there's short.

Microwaves and X-rays

You might take one look at this page and think it all looks fairly obvious — microwaves, used in microwave ovens and X-rays in X-ray machines, surely not. Ok, you probably knew that bit already, but this page does have lots of stuff about how they work, which isn't all that obvious, so don't miss it out.

Microwaves in *ovens* are *absorbed* by *water molecules*

1) The microwaves used in <u>microwave ovens</u> have a <u>different wavelength</u> to those used in communication.

2) These microwaves are actually <u>absorbed</u> by the water molecules in the food. They penetrate a few centimetres into the food before being <u>absorbed</u> by <u>water molecules</u>.

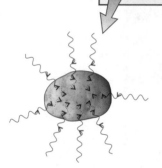

3) The energy is then <u>conducted</u> or <u>convected</u> to other parts.

4) If microwaves are absorbed by molecules in living tissue, <u>cells</u> may be <u>burned</u> or killed.

X-Rays are used to identify *fractures*

1) <u>Radiographers</u> in <u>hospitals</u> take <u>X-ray photographs</u> of people to see if they have any <u>broken bones</u>.

2) X-rays pass <u>easily through flesh</u> but not so easily through <u>denser material</u> like <u>bones</u> or <u>metal</u>. So it's the amount of radiation that's <u>absorbed</u> (or <u>not absorbed</u>) that gives you an X-ray image.

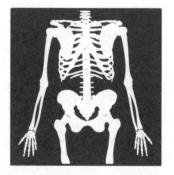

The <u>brighter bits</u> are where <u>fewer X-rays</u> get through. This is a <u>negative image</u>. The plate starts off <u>all white</u>.

3) X-rays can cause <u>cancer</u> (see page 240), so radiographers wear <u>lead aprons</u> and stand behind a <u>lead screen</u> or <u>leave the room</u> to keep their <u>exposure</u> to X-rays to a <u>minimum</u>.

Microwaves can be dangerous if they're absorbed by living tissue
X-rays are really useful for seeing inside people's bodies — but they can also cause cancer. So should we still use them? There's no right or wrong answer to questions like this — the best thing seems to be to only use x-rays when they're really needed and to try and protect the people who work with them.

More Uses of Waves

You've seen how electromagnetic waves can be dangerous, but they can also be really useful...

Infrared radiation can be used to monitor temperature

1) Infrared radiation (or IR) is also known as heat radiation. It's given out by all hot objects — and the hotter the object, the more IR radiation it gives out.

2) This means infrared can be used to monitor temperatures (see page 202).

3) Infrared is also detected by night-vision equipment:

Heat radiation is given off by all objects, even in the dark of night. The equipment turns it into an electrical signal, which is displayed on a screen as a picture.

Prenatal scanning uses ultrasound to make a picture

As the ultrasound hits different media some of the waves are partially reflected. These reflected waves are processed by computer to produce a video image of the foetus. No one knows for sure whether ultrasound is safe in all cases, but X-rays would definitely be dangerous to the foetus.

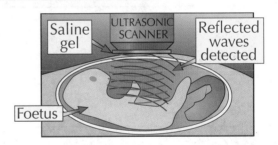

ADVANTAGES

1) As far as we know, it's safe.

2) It can show that the baby's alive and developing normally, determine its sex and show if it's likely to have Down's syndrome.

3) Ultrasound equipment is usually cheap and portable.

DISADVANTAGES

1) We can't be totally sure it's safe.

2) There are ethical issues — the parents may be more likely to want an abortion if the foetus is 'the wrong sex' or if it has Down's syndrome. Is it right for parents to have this choice?

It's a shame facts aren't as easily absorbed as infrared radiation

I know there's loads of uses of waves, but you really do need to know about them. Try writing a table with the type of wave, a use of it and a brief outline of how it works — it'll be a barrel of fun. Ok, maybe not, but it will be useful — you can even go back later and read it to get it in your head.

More Uses of Waves

Yes, there's more — waves are just great aren't they.

Iris scanning uses light to make a picture

1) Each eye's iris has a unique pattern — not even your own eyes have the same pattern as each other. This means that iris patterns can be used in security checks to prove a person's identity.

2) This is done using an iris scanner — which is basically like a camera. Light is reflected off your iris to make a picture. A computer then analyses the patterns in your iris to check your identity.

ADVANTAGES

1) There's very little chance of mistaking one iris code for another.

2) Your iris won't (usually) change during your life.

3) Iris recognition systems have more reference points for comparison than fingerprints.

4) Iris scanning is quick and easy to carry out.

DISADVANTAGES

1) Eye injuries or surgery (e.g. cataract removal) can occasionally change your iris pattern.

2) Some people feel their personal freedom is threatened: who's holding our data, and why?

3) If iris data is assumed to be 'unfakeable', will genuine mistakes or stolen identity be believed?

CD players use lasers to read information

1) The surface of a CD has a pattern of shallow pits (and higher areas called 'lands') cut into it.

2) A laser shone onto the CD is reflected from the shiny surface as it spins around in the player.

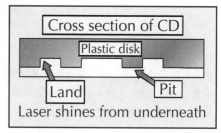

Cross section of CD
Plastic disk
Land   Pit
Laser shines from underneath

3) The beam is reflected from a land and a pit slightly differently — and this difference can be picked up by a light sensor. These differences in reflected signals can then be changed into an electrical signal.

4) An amplifier and a loudspeaker then convert the electrical signal into sound of the right pitch (frequency) and loudness.

Look into my eyes and tell me who I am

These are the last two uses of waves, I promise. Don't forget to add them to your table — you might want to add an advantages and disadvantages column too, so you can learn all about that.

Warm-Up and Exam Questions

There's a knack to passing exams — applying all the facts you've got stored in your brain to get as many marks as possible. These exam questions will help you to practise that.

Warm-up Questions

1) What happens when a wave is refracted, and why?
2) Explain how microwaves can heat food.
3) In what circumstances does total internal reflection of light occur?
4) What is an optical fibre?
5) Why don't optical fibres work if they're bent sharply?
6) Which type of wave is 'bounced back' by the ionosphere?

Exam Questions

1 The diagram shows a light ray entering and leaving a glass block.

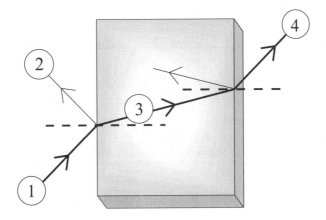

Match **A**, **B**, **C** and **D** to labels **1-4** on the diagram.

A A reflected ray

B A refracted ray

C An emergent ray

D An incident ray

(4 marks)

2 Which type of electromagnetic radiation can be detected by thermal imaging cameras?

A Radio waves

B Microwaves

C Infrared

D X-rays

(1 mark)

Exam Questions

3   The diagram shows a light ray entering an optical fibre.

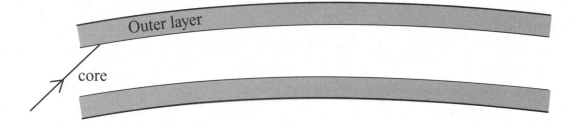

(a) Sketch the path taken by the light ray.

(1 mark)

(b) Explain why the light ray follows this path.

(1 mark)

4   The Government is planning to introduce identity cards containing biometric information such as iris patterns.

(a) Describe briefly how an iris scanner creates an image of an iris pattern.

(1 mark)

(b) Give one reason why checking iris patterns is a good way of identifying people.

(1 mark)

(c) Give one disadvantage of using iris recognition as an identification method.

(1 mark)

5   Jamie has an X-ray taken to see if he has broken his arm.

(a) Why are X-rays not very useful for looking at soft tissue injuries?

(1 mark)

(b) Give two ways in which radiographers can minimise their exposure to X-rays.

(2 marks)

6   Ultrasound is used to produce an image of the foetus in prenatal scans.

Give one advantage and one disadvantage of carrying out prenatal scans.

(2 marks)

Digital and Analogue Signals

Digital technology is gradually taking over. By 2012, you won't be able to watch TV unless you've got a digital version — that's when the Government's planning to switch off the last analogue signal.

Information is converted into signals

Information is being transmitted everywhere all the time.

1) Information, such as sounds and pictures, is converted into electrical signals before it's transmitted.

2) It's then sent long distances down telephone wires or carried on EM waves.

Analogue signals vary...

1) The amplitude and frequency of an analogue signal vary continuously. An analogue signal can take any value in a particular range.

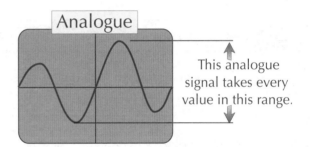

This analogue signal takes every value in this range.

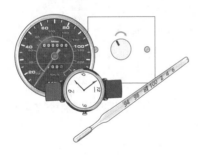

2) Dimmer switches, thermometers, speedometers and old-fashioned watches are all analogue devices.

...but digital's either on or off

1) Digital signals can only take two values (the two values sometimes get different names, but the key thing is that there are only two of them): on or off, true or false, 0 or 1...

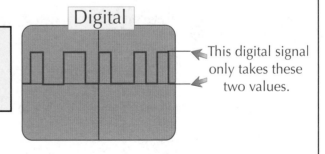

This digital signal only takes these two values.

2) On/off switches and the displays on digital clocks and meters are all digital devices.

Digital and Analogue Signals

It's all very well being able to transmit signals, but you need to be able to makes sense of them when you receive them too, or it's all a bit of a waste of time.

Signals have to be amplified

Both digital and analogue signals <u>weaken</u> as they travel, so they may need to be <u>amplified</u> along their route.

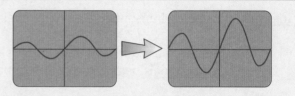

They also pick up <u>interference</u> or <u>noise</u> from <u>electrical disturbances</u> or <u>other signals</u>.

Digital signals are far better quality

1) <u>Noise</u> is less of a problem with <u>digital</u> signals.
   If you receive a noisy digital signal, it's pretty obvious what it's supposed to be.

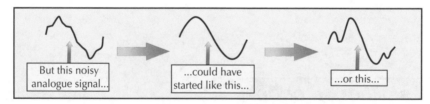

This noisy digital signal... ...is obviously supposed to be this.

2) But if you receive a noisy <u>analogue</u> signal, it's difficult to know what the <u>original</u> signal would have looked like. And if you amplify a noisy analogue signal, you amplify the noise as well.

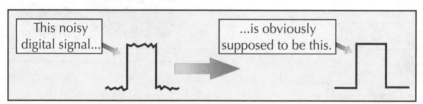

But this noisy analogue signal... ...could have started like this... ...or this...

3) This is why digital signals are much <u>higher quality</u> — the information received is the <u>same</u> as the original.

4) Digital signals are also easy to <u>process</u> using <u>computers</u>, since computers are digital devices too.

5) And another advantage of digital technology is that you can transmit <u>several signals at once</u> using just one cable or EM wave — so you can send <u>more information</u> (in a given time) than using analogue signals.

Remember: analogue varies but digital's either on or off

Digital signals are great — unless you live in a part of the country which currently has poor reception of digital broadcasts, in which case you get <u>no benefit at all</u>. This is because if you don't get spot-on reception of digital signals in your area, you won't get a grainy but watchable picture (like with analogue signals) — you'll get nothing at all. Except snow.

Seismic Waves

Seismic waves are completely different from EM waves (although they're still waves, so all the 'normal wave stuff' still applies). Seismic waves are produced by earthquakes.

Seismic waves are used to investigate Earth's structure

1) When there's an earthquake somewhere, it produces seismic waves which travel out through the Earth. We detect these waves all over the surface of the planet using seismographs.

2) There are two different types of seismic waves you need to learn — P-waves and S-waves:

3) Seismologists work out the time it takes for the shock waves to reach each seismograph. They also note which parts of the Earth don't receive the shock waves at all.

P-waves are longitudinal

1) P-waves travel through solids and liquids.

2) They travel faster than S-waves.

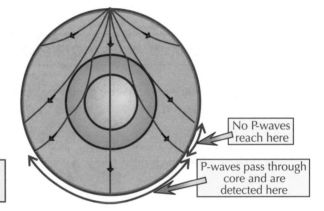

No P-waves reach here

P-waves pass through core and are detected here

S-waves are transverse

1) S-waves only travel through solids.

2) They're slower than P-waves.

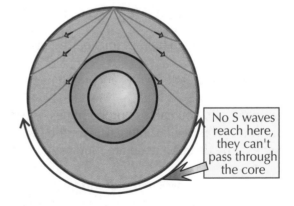

No S waves reach here, they can't pass through the core

You need to know the differences between S-waves and P-waves

Seismic waves might not seem earth-shakingly exciting to you, but those examiners love to ask about the differences between the two sorts, so you should try and learn them. It's not as hard as you might think, just remember to keep all the s-words together — s-waves, tranSverSe, solids, slow. It doesn't work quite as well for p-waves — but if you know what s-waves are you should be able to work it out.

Seismic Waves

The **seismograph** results tell us **what's down there**

1) About <u>halfway through</u> the Earth, both types of wave <u>change direction</u> abruptly. This indicates that there's a <u>sudden change</u> in <u>properties</u> — as you go from the <u>mantle</u> to the <u>core</u>.

2) <u>S-waves</u> do travel through the <u>mantle</u>, which shows that it's <u>solid</u>. But the fact that <u>S-waves</u> are <u>not detected</u> in the core's <u>shadow</u> tells us that the <u>outer core</u> is <u>liquid</u> — <u>S</u> waves only pass through <u>S</u>olids.

3) It's also found that <u>P-waves</u> travel <u>slightly faster</u> through the <u>middle</u> of the core, which strongly suggests that there's a <u>solid inner core</u>.

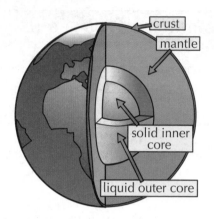

The waves **curve** with **increasing depth**

1) The <u>waves</u> change speed as the <u>properties</u> of the mantle and core change. This change in speed causes the waves to change direction — which is <u>refraction</u>, of course.

2) Most of the time the waves change speed <u>gradually</u>, resulting in a <u>curved path</u>. But when the properties change <u>suddenly</u>, the wave speed changes abruptly, and the path has a <u>kink</u>.

It's **difficult** to **predict earthquakes** and **tsunami waves**

<u>Some countries</u> are particularly <u>susceptible</u> to <u>earthquakes</u>, which can also sometimes <u>cause tsunami waves</u>. Both can be <u>extremely destructive</u>, especially in areas where the <u>housing</u> isn't built to <u>withstand</u> them. So it would be <u>very useful</u> to be able to <u>predict</u> when they're <u>likely to hit</u>.

1) Seismic waves can help predict earthquakes — a 'big one' usually happens after smaller 'foreshocks'.

2) But predicting earthquakes accurately is <u>hard</u>. Scientists don't agree on which method works best.

3) For example, you can use <u>probabilities</u> based on <u>previous occurrences</u> — e.g. if a city has had an earthquake at regular intervals over the last century, that <u>trend may continue</u>. This method isn't necessarily dead-on accurate, but it still gives the area <u>time to prepare</u>, just in case.

4) Other methods involve <u>monitoring</u> various factors such as the level of <u>groundwater</u>, <u>foreshocks</u>, the <u>emission of radon gas</u>, and even, in some countries, changes in <u>animal behaviour</u>. These methods have been <u>successful</u>, but have also <u>missed</u> some biggies.

Seismic waves are a good way of investigating the Earth's structure

Scientists are still doing lots of research into earthquake prediction — and disasters like the Indian Ocean tsunami in 2004 and the Kashmir earthquake in 2005 have really raised its profile of late.

Warm-Up and Exam Questions

Learning facts and practising exam questions is the only recipe for sure-fire success.
That's what the questions on this page are all about. All you have to do — is do them.

Warm-up Questions

1) Describe the difference between analogue and digital signals.
2) Which type of seismic wave travels faster — P-waves or S-waves?
3) Which type of seismic wave is longitudinal?
4) Why do seismic waves curve as they travel through the Earth?
5) Give one way in which scientists try to predict earthquakes.

Exam Questions

1   Communication signals pick up unwanted additional signals as they travel.
    This unwanted part of the signal is called

    A   analogue

    B   amplitude

    C   digital

    D   noise

                                                                        (1 mark)

2   When comparing P-waves and S-waves travelling through the Earth, which of the
    following statements is **not** true?

    A   They both travel through liquids.

    B   They both travel through solids.

    C   P-waves are faster than S-waves.

    D   They are both caused by earthquakes.

                                                                        (1 mark)

3   Rick is sick of not being able to get a clear signal on his analogue radio.
    He decides to replace it with a digital radio.

    (a) What values can digital signals take?

                                                                        (1 mark)

    (b) Explain why both digital and analogue radio signals need to be amplified when they
         reach his radio.

                                                                        (1 mark)

    (c) Give one reason why Rick will be able to hear the songs on the radio more clearly
         with his digital radio than with his analogue radio.

                                                                        (1 mark)

Revision Summary for Section Eight

Try these lovely questions. Go on — you know you want to. It'll be nice.

1) Explain what current, voltage and resistance are in an electric circuit.

2) Do batteries produce AC or DC current? In what units is battery capacity measured?

3) Sketch typical voltage-current graphs for: a) a resistor, and b) a filament lamp.
    Explain the shape of each graph.

4)* Calculate the resistance of a wire if the voltage across it is 12 V and the current through it is 2.5 A.

5) Describe how the resistance of an LDR varies with light intensity. Give an application of an LDR.

6) Describe how you can create a current in a coil of wire using a magnet.

7) What are three factors that affect the size of the voltage that you get this way?

8) Sketch a generator with all the details. Explain how it works.

9) Sketch a typical power station, and explain what happens at each stage. Describe the useful energy
    transformations that occur.

10) Explain why a very high electrical voltage is used to transmit electricity in the national grid.

11) What's the name of the type of transformer that increases voltage? Where are these used?

12) Write down the formula linking voltage, current and power.

13)*Calculate the energy used by a 3.1 kW kettle if it's on for 1½ minutes.

14)*a) How many units of electricity (in kWh) would a kettle of power 2500 W use in 2 minutes?
    b) How much would that cost, if one unit of electricity costs 12p?

15) What is an 'RCCB' and how does it prevent electrocution?

16) Draw a diagram of a wave and label a crest and a trough, and the wavelength and amplitude.

17) Sketch the EM spectrum with all the details you've learned. Put the lowest frequency waves first.

18) Electromagnetic waves don't carry any matter. What <u>do</u> they carry?

19) What aspect of EM waves determines their differing properties?

20)*Find the speed of a wave with frequency 50 kHz and wavelength 0.3 cm.

21) Describe what can happen to a wave's direction when it meets an obstacle.

22) Draw a diagram showing the diffraction of a wave as it passes through: a) a small gap, b) a big gap.

23) Describe the main <u>known</u> dangers of microwaves, infrared, visible light, UV and X-rays.

24) What has led to a thinning of the ozone layer? Why is this a problem for humans?

25) Explain what is meant by: a) refraction b) total internal reflection.

26) Explain why sending data by optical fibre might be better than broadcasting it as a radio signal.

27)*In which of the cases A to D below would the ray of light be totally internally reflected?
    (The critical angle for glass is approximately 42°.)

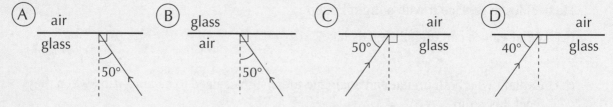

28) Describe the different ways that short, medium and long-wave radio signals can travel long distances.

29) Describe two uses of microwaves, and explain why microwaves are suitable for these uses.

30) Explain how EM waves are used in: a) prenatal scanning, b) iris scanning, c) CD players.

31) Draw diagrams illustrating analogue and digital signals. What advantages do digital signals have?

32) How do P-waves and S-waves differ regarding: a) type of wave, b) speed, c) what they go through?

* Answers on page 296.

Radioactivity

Nuclei contain protons and neutrons

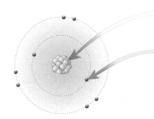

The nucleus contains protons and neutrons. It makes up most of the mass of the atom, but takes up virtually no space — it's tiny.

The electrons are negatively charged and really really small.

They whizz around the outside of the atom. Their paths take up a lot of space, giving the atom its overall size (though it's mostly empty space).

Isotopes are atoms with different numbers of neutrons

1) Many elements have a few different isotopes. Isotopes are atoms with the same number of protons but a different number of neutrons.

2) E.g. there are two common isotopes of carbon. The carbon-14 isotope has two more neutrons than 'normal' carbon (carbon-12).

3) Usually each element only has one or two stable isotopes — like carbon-12. The other isotopes tend to be radioactive — the nucleus is unstable, so it decays (breaks down) and emits radiation. Carbon-14 is an unstable isotope of carbon.

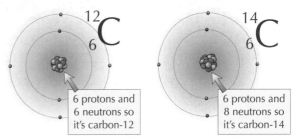

$^{12}_{6}C$

6 protons and 6 neutrons so it's carbon-12

$^{14}_{6}C$

6 protons and 8 neutrons so it's carbon-14

Radioactive decay is a random process

1) The nuclei of unstable isotopes break down at random. If you have 100 unstable nuclei, you can't say when any one of them is going to decay, and you can't do anything to make a decay happen.

2) Each nucleus just decays quite spontaneously in its own good time. It's completely unaffected by physical conditions like temperature or any sort of chemical bonding etc.

3) When the nucleus does decay it spits out one or more of the three types of radiation — alpha, beta and gamma (see next page).

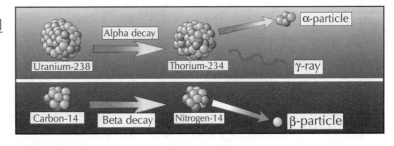

Uranium-238   Alpha decay   Thorium-234   α-particle   γ-ray

Carbon-14   Beta decay   Nitrogen-14   β-particle

4) In the process, the nucleus often changes into a new element.

Background radiation is everywhere all the time

There's (low-level) background nuclear radiation all around us all the time. It comes from:

• substances here on Earth — some radioactivity comes from air, food, building materials, soil, rocks...

• radiation from space (cosmic rays) — these come mostly from the Sun (see page 270),

• living things — there's a little bit of radioactive material in all living things,

• radiation due to human activity — e.g. fallout from nuclear explosions, or nuclear waste (though this is usually a tiny proportion of the total background radiation).

Three Kinds of Radioactivity

There are three types of radiation — alpha (α), beta (β) (on this page) and gamma (γ) (on the next page). You need to remember <u>what</u> they are, how well they <u>penetrate</u> materials (including air), and their <u>ionising</u> power.

Nuclear radiation causes ionisation

1) Nuclear radiation causes <u>ionisation</u> by <u>bashing into atoms</u> and <u>knocking electrons off</u> them. Atoms (with <u>no overall charge</u>) are turned into <u>ions</u> (which are <u>charged</u>) — hence the term "ionisation".

2) There's a pattern: the <u>further</u> the radiation can <u>penetrate</u> before hitting an atom and getting stopped, the <u>less damage</u> it will do along the way and so the <u>less ionising</u> it is.

Alpha particles are helium nuclei

1) Alpha particles are made up of <u>2 protons and 2 neutrons</u> — they're <u>big</u>, <u>heavy</u> and <u>slow-moving</u>.

2) They therefore <u>don't penetrate</u> far into materials but are <u>stopped quickly</u>.

3) Because of their size they <u>bash into a lot of atoms</u> and <u>knock electrons off</u> them before they slow down, which creates lots of ions.

4) Because they're electrically <u>charged</u> (with a positive charge), alpha particles are <u>deflected</u> (their <u>direction changes</u>) by <u>electric</u> and <u>magnetic fields</u>.

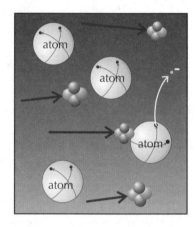

Beta particles are electrons

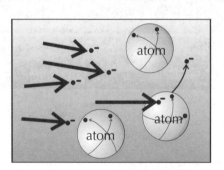

1) A beta particle is an <u>electron</u> which has been emitted from the <u>nucleus</u> of an atom when a <u>neutron</u> turns into a <u>proton</u> and an <u>electron</u>. So for every β-particle emitted, the number of <u>protons</u> in the nucleus increases by 1.

2) They move <u>quite fast</u> and they are <u>quite small</u>.

3) They <u>penetrate moderately</u> before colliding and are <u>moderately ionising</u> too.

4) Because they're <u>charged</u> (negatively), beta particles are <u>deflected</u> by electric and magnetic fields.

Learning the types of radiation is as easy as α, β, γ

The symbols for alpha, beta and gamma radiation may look a little strange — but really they're just a, b and c written using the Greek alphabet. True it might be easier to use a, b and c, but the Greek letters have been used for so long now that it'd confuse more people than it would help, sorry.

Three Kinds of Radioactivity

Gamma radiation is quite different from alpha and beta radiation. Gamma rays are part of the electromagnetic spectrum — just like light and radio waves.

Gamma rays are very short wavelength EM waves

1) In a way, gamma rays are the opposite of alpha particles. They have no mass — they're just energy (in the form of an EM wave — see page 236).

2) They penetrate a long way into materials without being stopped.

3) This means they are weakly ionising because they tend to pass through rather than collide with atoms. But eventually they hit something and do damage.

4) Gamma rays have no charge, so they're not deflected by electric or magnetic fields.

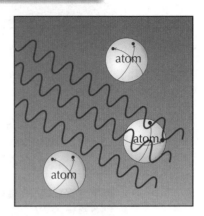

You can identify the type by what blocks it

Make sure you know what it takes to block each of the three types of radiation:

Alpha particles are blocked by paper, skin or a few centimetres of air.

Beta particles are stopped by thin metal.

Gamma rays are blocked by thick lead or very thick concrete.

| Thin mica | Skin or paper stops ALPHA | Thin aluminium stops BETA | Thick lead stops GAMMA |

So if radiation can penetrate paper it could be beta or gamma — you'd have to test it with a metal, say, to find out which.

Remember — alpha penetrates least, gamma penetrates most

Remember: alpha's big, slow and clumsy — always knocking into things. Beta's lightweight and fast, and gamma weighs nothing and moves super-fast. Practise with this: if it gets through paper and is deflected by a magnetic field, it must be _____ radiation. (Answer on page 296.)

Half-Life

The **radioactivity** of a sample always **decreases over time**

1) Each time an unstable nucleus <u>decays</u> and emits radiation, that means one more <u>radioactive</u> <u>nucleus</u> <u>isn't there</u> to decay later.

2) As more <u>unstable nuclei</u> decay, the <u>radioactivity</u> of the source as a whole <u>decreases</u> — so the <u>older</u> a radioactive source is, the <u>less radiation</u> it emits.

3) <u>How quickly</u> the activity <u>decreases</u> varies a lot. For <u>some</u> isotopes it takes <u>just a few hours</u> before nearly all the unstable nuclei have <u>decayed</u>. Others last for <u>millions of years</u>.

4) The problem with trying to <u>measure</u> this is that <u>the activity never reaches zero</u>, which is why we have to use the idea of <u>half-life</u> to measure <u>how quickly the activity decreases</u>.

5) Learn this <u>important definition</u> of <u>half-life</u>:

> *<u>HALF-LIFE</u> is the <u>TIME TAKEN</u> for <u>HALF</u> of the <u>nuclei</u> now present to <u>DECAY</u>*

6) A <u>short half-life</u> means the <u>activity falls quickly</u>, because <u>lots</u> of the nuclei decay <u>quickly</u>.

7) A <u>long half-life</u> means the activity <u>falls more slowly</u> because <u>most</u> of the nuclei don't decay <u>for a long time</u> — they just sit there, <u>basically unstable</u>, but kind of <u>biding their time</u>.

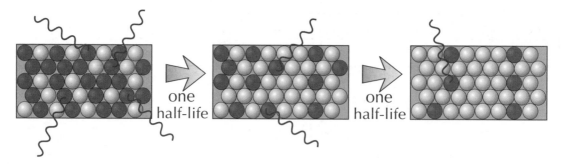

Half-life is the **time taken** for the **count rate** to **halve**

You can work out the <u>half-life</u> of a sample by monitoring its <u>count rate</u> (the number of atoms which decay per minute) using a <u>Geiger counter</u>. Then plot a graph of <u>count rate</u> against <u>time</u>.

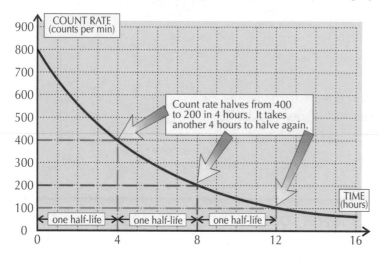

Count rate halves from 400 to 200 in 4 hours. It takes another 4 hours to halve again.

1) This is how 'carbon dating' of some types of archaeological specimens works...

2) If you know the <u>half-life</u> of a particular radioactive isotope you can look at <u>how much is left</u> in your specimen now, and work out <u>how long</u> your dusty bits of bone have been around for.

3) This method is also really important in <u>geology</u> and <u>biology</u> — for working out the age of rocks (and of any <u>fossils</u> that are buried in them).

Warm-Up and Exam Questions

There's no point in skimming through the section and glancing over the questions. Do the warm-up questions and go back over any bits you don't know. Then try the exam questions — without cheating.

Warm-up Questions

1) What is meant by 'radioactive decay'?
2) Explain what isotopes are.
3) Name the three types of nuclear radiation.
4) Which type of nuclear radiation is also a type of electromagnetic radiation?
5) Give one source of background radiation.

Exam Questions

1 The diagram shows four different materials.

thin mica	skin	aluminium	thick lead
1	2	3	4

Match up the materials **1-4** with these descriptions.

A Stops all types of nuclear radiation.

B Doesn't stop any types of nuclear radiation.

C Stops alpha and beta radiation.

D Stops only alpha radiation.

(4 marks)

2 An alpha particle is

A a proton

B a neutron

C a helium nucleus

D an electromagnetic wave

(1 mark)

3 A sample of a highly ionising radioactive gas has a half-life of two minutes.

(a) What does 'half-life' mean?

(1 mark)

(b) What fraction of the radioactive atoms currently present will be left after four minutes?

(1 mark)

(c) When an atom of the gas decays, it releases an electron.
What type of nuclear radiation does this gas emit?

(1 mark)

Dangers from Nuclear Radiation

Nuclear radiation can do nasty stuff to living cells so you have to be careful how you handle it.
The effect of radiation on cells depends on the type of radiation and the size of the dose.

Radiation harms living cells...

1) <u>Beta</u> and <u>gamma</u> radiation can penetrate the skin
and soft tissues to reach the delicate <u>organs</u> inside
the body. This makes beta and gamma sources
more hazardous than alpha when outside the body.

2) Alpha radiation can't penetrate the skin, but it's a
different story when it gets inside your body (by
<u>swallowing</u> or <u>breathing it in</u>, say) — alpha sources
do all their damage in a <u>very localised area</u>.

3) Beta and gamma sources, however, are <u>less
dangerous</u> inside the body — their radiation mostly
<u>passes straight out</u> without doing as much damage.

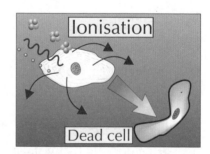

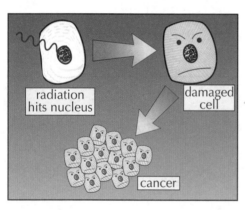

4) If radiation enters your body, it will <u>collide</u>
with molecules in your cells.

5) These collisions cause <u>ionisation</u>, which
<u>damages</u> or <u>destroys</u> the molecules.

6) <u>Lower</u> doses tend to cause <u>minor</u> damage without
<u>killing</u> the cell. This can give rise to <u>mutant</u> cells
which <u>divide uncontrollably</u> — this is <u>cancer</u>.

7) <u>Higher</u> doses tend to <u>kill cells</u> completely, causing
<u>radiation sickness</u> if a large part of your body is
affected at the same time.

8) The <u>extent</u> of the harmful effects depends on
<u>how much exposure</u> you have to the radiation,
and its <u>energy</u> and <u>penetration</u>.

Nuclear radiation + living cells = cell damage, cell death or cancer

Most people could probably tell you that nuclear radiation is dangerous — what you need to know is
what radiation can do to living cells and why the three types of radiation have different effects. Check
out pages 258-259 if you're having trouble remembering the properties of the different radiation types.

Dangers from Nuclear Radiation

You've got to be really careful with anything radioactive — no mucking about.

You should **protect yourself** in the **laboratory**...

You should always act to <u>minimise</u> your exposure to radioactive sources.

1) <u>Never</u> allow <u>skin contact</u> with a source. Always handle sources with <u>tongs</u>.

2) Keep the source at <u>arm's length</u> to keep it <u>as far</u> from the body <u>as possible</u>.

3) Keep the source <u>pointing away</u> from the body and <u>avoid looking directly at it</u>.

4) <u>Always</u> keep the source in a <u>labelled lead box</u> and put it back in <u>as soon</u> as the experiment is <u>over</u>, to keep your <u>exposure time</u> short.

...and if you **work** with **nuclear radiation**

1) Industrial nuclear workers wear <u>full protective suits</u> to prevent <u>tiny radioactive particles</u> being <u>inhaled</u> or lodging <u>on the skin</u> or <u>under fingernails</u> etc.

2) <u>Lead-lined suits</u> and <u>lead/concrete barriers</u> and <u>thick lead screens</u> are used to prevent exposure to <u>gamma rays</u> from highly contaminated areas. (α and β radiation are stopped <u>much more easily</u>.)

3) Workers use <u>remote-controlled robot arms</u> to carry out tasks in highly radioactive areas.

Radiation's dangerous stuff — safety precautions are crucial

It's quite difficult to do research on how radiation affects humans. This is partly because it would be <u>unethical</u> to do <u>controlled experiments</u>, exposing people to huge doses of radiation just to see what happens. We rely mostly on studies of populations affected by <u>nuclear accidents</u> or nuclear <u>bombs</u>.

Uses of Nuclear Radiation

Nuclear radiation can be very <u>dangerous</u>. But it can be very <u>useful</u> too. Read on...

Alpha radiation is used in smoke detectors

1) Smoke detectors have a <u>weak</u> source of <u>α-radiation</u> close to <u>two electrodes</u>.

2) The radiation <u>ionises</u> the air, and a <u>current</u> flows between the electrodes.

3) But if there's a fire, the <u>smoke</u> <u>absorbs</u> the <u>radiation</u> — the <u>current stops</u> and the <u>alarm sounds</u>.

Beta radiation is used in tracers and thickness gauges

Medical tracers...

For example, if a radioactive source is <u>injected</u> into a patient (or <u>swallowed</u>), its progress around the body can be followed using an <u>external radiation detector</u>.

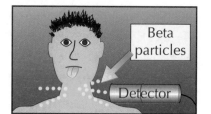

A computer converts the reading to a TV display showing where the strongest reading is coming from.

These '<u>tracers</u>' can show if the body is working properly.

Doctors use <u>beta</u> or <u>gamma</u> emitters as <u>tracers</u> because this radiation <u>passes out</u> of the body. They also choose things that are only radioactive for a <u>few hours</u>.

Thickness control...

1) You direct radiation through the stuff being made (e.g. paper or cardboard), and put a detector on the other side, connected to a control unit.

2) When the amount of <u>detected</u> radiation goes <u>down</u>, it means the paper is coming out <u>too thick</u>, so the control unit pinches the rollers up a bit to make it thinner.

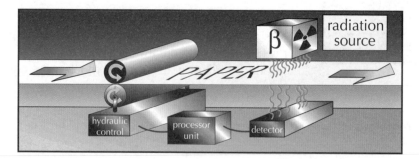

3) If the reading goes <u>up</u>, the paper's <u>too thin</u>, so the control unit opens the rollers out a bit.

For this use, your radioactive substance mustn't decay away <u>too quickly</u>, otherwise its strength would gradually fall (and the control unit would keep pinching up the rollers trying to compensate).

You need to use a <u>beta</u> source, because then the paper or cardboard will <u>partly block</u> the radiation. If it <u>all</u> goes through (or <u>none</u> of it does), then the reading <u>won't change</u> at all as the thickness changes.

Uses of Nuclear Radiation

Gamma radiation has medical and industrial uses

Treating cancer...

1) High doses of <u>gamma rays</u> will kill living cells, so they can be used to treat <u>cancers</u>.

2) The gamma rays have to be <u>directed carefully</u> at the cancer, and at just the right <u>dosage</u> so as to kill the <u>cancer</u> cells <u>without</u> damaging too many <u>normal</u> cells.

Sterilising medical instruments...

1) Gamma rays are also used to <u>sterilise</u> medical instruments — by <u>killing</u> all the microbes.

2) This is better than trying to <u>boil</u> plastic instruments, which might be damaged by high temperatures.

3) You need to use a strongly radioactive source that lasts a long time, so that it doesn't need replacing too often.

Non-destructive testing...

Several industries also use gamma radiation to do <u>non-destructive testing</u>.

For example, <u>airlines</u> can check the turbine blades of their jet engines by directing gamma rays at them — if too much radiation <u>gets through</u> the blade to the <u>detector</u> on the other side, they know the blade's <u>cracked</u> or there's a fault in the welding.

 It's so much better to find this out before you take off than in mid-air.

The isotope you use depends on half-life and whether it's α, β or γ

Knowing the detail is important here. For instance, swallowing an alpha source as a medical tracer would be very foolish — alpha radiation would cause all sorts of chaos inside your body but couldn't be detected outside, making the whole thing pointless. So learn <u>what</u> each type's used for <u>and why</u>.

Warm-Up and Exam Questions

Imagine if you opened up your exam paper and all the answers were already written in for you.
Hmm, well I'm afraid that won't happen, so the only way you'll do well is through some hard work now.

Warm-up Questions

1) How do smoke detectors work?
2) Why is nuclear radiation dangerous?
3) Describe two precautions that should be taken when handling radioactive sources in the lab.
4) Give one way in which workers in nuclear power plants can be protected from radiation.
5) Give one example of a medical use of nuclear radiation.

Exam Questions

1   Which type of nuclear radiation is the most dangerous inside the body?

    A   alpha

    B   beta

    C   gamma

    D   neutron

(1 mark)

2   Temperature-sensitive surgical instruments are sterilised by

    A   alpha radiation

    B   boiling

    C   beta radiation

    D   gamma radiation

(1 mark)

3   The diagram shows paper being made in a mill.

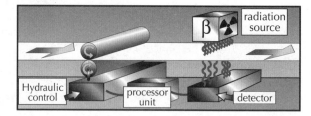

    (a)   Describe how the thickness of the paper is controlled using beta radiation.

(2 marks)

    (b)   Why isn't an alpha source used with this machinery?

(1 mark)

    (c)   Why isn't gamma radiation used for this purpose?

(1 mark)

The Solar System

Our Solar System consists of a star (the Sun) and lots of stuff orbiting it in elongated circles.

There are **eight planets** in our Solar System

- Closest to the Sun are the <u>inner planets</u> — Mercury, Venus, Earth and Mars.
- Then the <u>asteroid belt</u> — see below.
- Then the <u>outer planets</u>, much further away — Jupiter, Saturn, Uranus, Neptune.

We used to say there was a ninth planet called Pluto, but scientists have decided it's too small to be counted as a planet. It's now called 'minor planet 134340 Pluto' — catchy.

You need to learn the <u>order</u> of the planets, which is made easier by using the little jollyism below:

Mercury,	Venus,	Earth,	Mars,	(Asteroids),	Jupiter,	Saturn,	Uranus,	Neptune
(Mad	Vampires	Eat	Mangos	And	Jump	Straight	Up	Noses)

Planets **reflect sunlight** and orbit the **Sun** in **ellipses**

1) You can <u>see</u> some planets with the <u>naked eye</u> — they look like <u>stars</u>, but they're <u>totally different</u>.

2) Stars are <u>huge</u>, very <u>hot</u> and very <u>far away</u>. They <u>give out</u> lots of <u>light</u> — which is why you can see them, even though they're very far away.

3) The planets are <u>smaller</u> and <u>nearer</u> and they just <u>reflect sunlight</u> falling on them.

4) Planets often have <u>moons</u> orbiting around them. Jupiter has at least 63 of 'em. We've just got one.

There's a **belt** of **asteroids** orbiting between **Mars** and **Jupiter**

1) When the Solar System formed, there were lots of leftovers — <u>bits of rock and rubble</u>. These included <u>asteroids</u>, which orbit the Sun in the <u>asteroid belt</u>, between the orbits of <u>Jupiter</u> and <u>Mars</u>. (They didn't get to form a planet because the large gravitational force of Jupiter kept interfering.)

2) Occasionally lumps of rock get <u>pushed out</u> of their <u>nice safe orbits</u> (e.g. due to a collision) — they might then enter the Earth's atmosphere...

3) ...where they're called <u>meteors</u>. As they pass through our <u>atmosphere</u> they usually <u>burn up</u> and we see them as <u>shooting stars</u>.

4) Sometimes, not all of the meteor burns up and part crashes into the <u>Earth's surface</u> as a <u>meteorite</u>.

The Solar System

Every now and then an asteroid or comet comes near the Earth — scary stuff. Luckily, the chances of an object hitting the Earth and being big enough to do serious damage is really low, phew.

Comets orbit the Sun in very elliptical orbits

1) Comets are balls of ice and dust which orbit the Sun in very elongated ellipses — taking them very far away and then back in close — which is when we see them.

2) The Sun is near one end of the orbit, not at its centre.

3) As a comet approaches the Sun, it speeds up. Also, its ice melts, leaving a bright tail of gas and debris which can be millions of kilometres long.

Comet

Near-Earth Objects (NEOs) could collide with Earth

1) Some asteroids and comets have orbits that pass very close to the orbit of the Earth. [They're called Near-Earth Objects (NEOs).]

2) Astronomers use powerful telescopes and satellites to search for and monitor NEOs. Then they can calculate an object's trajectory (the path it's going to take) and find out if it's heading for us.

3) A huge collision is unlikely in our lifetimes, but it's possible — there have certainly been big impacts in the past. Evidence for this includes:

> i) big craters,

> ii) layers of unusual elements in rocks — these must have been 'imported' by an asteroid,

> iii) sudden changes in fossil numbers between adjacent layers of rock, as species suffer extinction.

4) If we get enough warning, we could try to deflect an NEO before it hits us. Scientists have various ideas about this — you could explode a nuclear bomb next to the object to 'nudge' it off course, or you could speed it up (or slow it down) so that it reaches Earth's orbit when we're out of the way.

Don't get asteroids, meteors and comets confused

It's serious business, this NEO stuff. In 2002, an asteroid narrowly missed us, and we only found it 12 days beforehand — not very much time to put a plan together. Even if Bruce Willis had been on hand.

Magnetic Fields and Solar Flares

The Earth can protect us from a lot of nasties from space — e.g. cosmic rays.

The Earth is a bit like a big magnet

1) The Earth is surrounded by a <u>magnetic field</u> — a region where <u>magnetic materials</u> (like iron and steel) experience a <u>force</u>. Basically, this means the Earth acts like a big <u>bar magnet</u>.

2) Like all magnets, it has <u>north</u> and <u>south poles</u>.

> But... (concentrate now) the Earth's <u>south magnetic pole</u> is actually at the <u>North Pole</u>. This makes sense if you think about it... if you have a <u>compass</u> (or any other magnet), its north pole <u>points north</u> — because it's attracted towards a <u>south magnetic pole</u> (remember, opposites attract).

3) You can use a compass to tell the direction of the magnetic field. The needle points in the direction of the field.

4) The Earth <u>doesn't</u> actually have a <u>giant bar magnet</u> buried inside it. Its <u>core</u> contains a lot of <u>molten iron</u> and <u>nickel</u>, which move in <u>convection currents</u>. Scientists don't understand this fully, but they think that <u>electric currents</u> within this <u>liquid core</u> create the <u>magnetic field</u> (and vice versa).

There's more to the Earth than meets the eye

The Earth acting like a giant magnet sounds a bit funny at first — but it makes sense when you think about it. Remember, the Earth's south magnetic pole is at the North pole and vice versa. Just think of a compass — the needle's north pole points North, so it must be pointing at a south magnetic pole.

Magnetic Fields and Solar Flares

Earth's magnetic field **shields** us from **charged particles**

1) The surface of the Sun is a very unpleasant place — the Sun's constantly releasing enormous amounts of <u>energy</u> and <u>cosmic rays</u>. Cosmic rays are heavily <u>ionising</u>, and mostly consist of <u>charged particles</u>, though there's gamma rays and X-rays in there too.

2) The Earth's magnetic field does a good job of <u>shielding</u> us from a lot of the charged particles from the Sun by deflecting them away. But when charged particles in <u>cosmic rays</u> do hit the Earth's atmosphere they create <u>gamma rays</u>, forming part of the Earth's background radiation — see page 257.

DETLEV VAN RAVENSWAAY/
SCIENCE PHOTO LIBRARY

3) From time to time, massive <u>explosions</u> called <u>solar flares</u> also occur on the surface of the Sun. Solar flares release <u>vast</u> amounts of energy, some of it as <u>gamma rays</u> and <u>X-rays</u>.

4) Solar flares also give off <u>massive</u> clouds of <u>charged particles</u>. Some reach the Earth and produce <u>disturbances</u> in the Earth's <u>magnetic field</u>.

5) Solar flares can also damage <u>artificial satellites</u>, which we rely on for <u>all sorts</u> of things — for example... modern <u>communications</u>, <u>weather forecasting</u>, <u>spying</u>, <u>navigation</u> systems (such as GPS)...

6) The problem is that <u>electrons</u> and <u>ions</u> in solar flares can cause <u>surges</u> of <u>current</u> in a satellite's electrical circuitry. So satellites might need to be <u>shut down</u> to prevent damage during flares.

7) Solar flares also interact with the Earth's magnetic field, and can cause <u>power surges</u> in electricity distribution systems here on Earth (by means of <u>electromagnetic induction</u> — see page 224).

Cosmic rays cause the Aurora Borealis

1) Some charged particles in cosmic rays are <u>deflected</u> by the Earth's magnetic field and spiral down near the <u>magnetic poles</u>. Here, some of their <u>energy</u> is transferred to particles in the Earth's atmosphere, causing them to emit light — the <u>polar lights</u>.

2) The polar lights are shifting 'curtains' of light that appear in the sky. They're called the aurora borealis (<u>northern lights</u>) at the North Pole and the aurora australis at the South Pole. These displays are more dramatic during solar flares, when more cosmic rays arrive at the Earth.

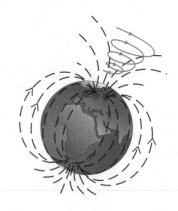

Solar flares can cause some big problems

In an old Scandinavian language, the word for 'northern lights' is 'herring flash' — people used to think the lights were <u>reflections</u> cast into the sky by large shoals of <u>herring</u>. That's not an <u>entirely</u> stupid idea, but <u>not</u> what you call the <u>accepted explanation</u> nowadays — so don't try it in the exam.

Beyond the Solar System

There's all sorts of exciting stuff in the Universe... Our whole Solar System is just part of a huge galaxy. And there are billions upon billions of galaxies. You should be realising now that the Universe is huge...

We're in the **Milky Way galaxy**

1) Our Sun is just one of many billions of stars which form the Milky Way galaxy. Our Sun is about halfway along one of the spiral arms of the Milky Way.

2) The distance between neighbouring stars in the galaxy is usually millions of times greater than the distance between planets in our Solar System.

3) The force which keeps the stars together in a galaxy is gravity, of course. And like most things in the Universe, galaxies rotate — a bit like a Catherine wheel.

The **whole universe** has **more than a billion galaxies**

1) Galaxies themselves are often millions of times further apart than the stars are within a galaxy.

2) So even the slowest among you will have worked out that the Universe is mostly empty space and is really really BIG.

Scientists measure distances in **space** using **light years**

1) Once you get outside our Solar System, the distances between stars and between galaxies are so enormous that kilometres seem too pathetically small for measuring them.

2) For example, the closest star to us (after the Sun) is about 40 000 000 000 000 kilometres away (give or take a few hundred billion kilometres). Numbers like that soon get out of hand.

3) So scientists use light years instead — a light year is the distance that light travels through a vacuum (like space) in one year. Simple as that.

4) If you work it out, 1 light year is equal to about 9 460 000 000 000 kilometres. Which means the closest star after the Sun is about 4.2 light years away from us.

5) Just remember — a light year is a measure of DISTANCE (not time).

Stars can **explode** — and they sometimes leave **black holes**

1) When a really big star has used up all its fuel, it explodes. What's left afterwards is really dense — sometimes so dense that nothing can escape its gravitational pull. It's now called a black hole.

2) Black holes have a very large mass but their diameter is tiny in comparison.

3) They're not visible — even light can't escape their gravitational pull (that's why it's 'black').

4) Astronomers have to detect black holes in other ways — e.g. they can observe X-rays emitted by hot gases from other stars as they spiral into the black hole.

Warm-Up and Exam Questions

Here's some nice warm-up questions to get you into the swing of things before you try the exam questions.

Warm-up Questions

1) Between which two planets in the Solar System are most asteroids found?
2) What are comets made of?
3) Give one difference between a comet and an asteroid.
4) What is a galaxy?
5) Why is a black hole black?

Exam Questions

1   The diagram shows a simple picture of part of the
     inner Solar System (not to scale).
     Match objects **1-4** with the descriptions given below.

     A   Venus

     B   Earth

     C   Mars

     D   comet

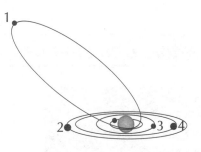

(4 marks)

2   A light year is the

     A   size of a galaxy

     B   time taken for light to travel from the Sun to the Earth

     C   distance light travels in one year

     D   speed of the nearest galaxy to our own

(1 mark)

3   Solar flares occur on the surface of the Sun.

     (a)   What are solar flares?

(1 mark)

     (b)   (i)   What shields the Earth from most of the charged particles in solar flares?

(1 mark)

           (ii) What effect can often be seen from near the North Pole as charged particles enter
                 the atmosphere?

(1 mark)

     (c)   (i)   Give one use of artificial satellites.

(1 mark)

           (ii) How can solar flares cause problems for artificial satellites?

(1 mark)

The Life Cycle of Stars

Stars go through many traumatic stages in their lives — just like teenagers.

Clouds of Dust and Gas

1) Stars initially form from clouds of DUST AND GAS.

Protostar

2) The force of gravity makes the gas and dust spiral in together to form a protostar. Gravitational energy has been converted into heat energy, so the temperature rises.

Main Sequence Star

3) When the temperature gets high enough, hydrogen nuclei undergo nuclear fusion to form helium nuclei and give out massive amounts of heat and light. A star is born. It immediately enters a long stable period where the heat created by the nuclear fusion provides an outward pressure to balance the force of gravity pulling everything inwards. In this stable period it's called a MAIN SEQUENCE STAR and it lasts several billion years. (The Sun is in the middle of this stable period — or to put it another way, the Earth has already had half its innings before the Sun engulfs it!)

Red Giant

4) Eventually the hydrogen begins to run out and the star then swells into a RED GIANT. It becomes red because the surface cools.

5) A small star like our Sun will then begin to cool and contract into a WHITE DWARF and then finally, as the light fades completely, it becomes a BLACK DWARF. (That's going to be really sad.)

Small stars

Big stars

White Dwarf

Black Dwarf

6) Big stars, however, start to glow brightly again as they undergo more fusion and expand and contract several times, forming heavier elements in various nuclear reactions. Eventually they'll explode in a SUPERNOVA.

new planetary nebula...   ...and a new solar system

Supernova

Neutron Star...

7) The exploding supernova throws the outer layers of dust and gas into space, leaving a very dense core called a NEUTRON STAR. If the star is big enough this will become a BLACK HOLE (see page 271).

...or Black Hole

8) The dust and gas thrown off by the supernova will form into SECOND GENERATION STARS like our Sun. The heavier elements are only made in the final stages of a big star just before and during the final supernova, so the presence of heavier elements in the Sun and the inner planets is clear evidence that our beautiful and wonderful world, with its warm sunsets and fresh morning dews, has all formed out of the snotty remains of a grisly old star's last dying sneeze.

What a star's like during its life determines what it becomes when it's dead

Erm. Now how do they know that exactly... Anyway, now you know what the future holds — our Sun is going to fizzle out, and it'll just get very very cold and very very dark. Great.

The Origins of the Universe

How did it all begin... Well, once upon a time, there was a really <u>Big Bang</u> — that's the <u>most convincing theory</u> we've got for how the Universe started.

The **Universe** seems to be **expanding**

As big as the Universe is, it looks like it's getting even bigger. All its galaxies seem to be moving away from each other. There's good evidence for this...

Light from other galaxies is red-shifted

1) When we look at <u>light from distant galaxies</u> we find that the <u>frequencies</u> are all <u>lower</u> than they should be — they're <u>shifted</u> towards the <u>red end</u> of the spectrum.

2) This is called the <u>red-shift</u>. (It's the same effect as a car's noise sounding <u>lower-pitched</u> when the car's <u>moving away</u> from you.)

3) <u>Measurements</u> of red-shift suggest that <u>all the galaxies</u> are <u>moving away from us</u> very quickly — and it's the <u>same result</u> whichever direction you look in.

More distant galaxies have greater red-shifts

1) <u>More distant</u> galaxies have <u>greater</u> red-shifts than nearer ones.

2) This means that more distant galaxies are <u>moving away faster</u> than nearer ones. All these findings indicate that the whole Universe is <u>expanding</u>.

There's **uniform microwave radiation** from all **directions**

This is another observation that scientists have made. It's not all that interesting in itself, but the theory that explains all this evidence definitely is.

1) Scientists have detected <u>low frequency electromagnetic radiation</u> coming from <u>all parts</u> of the Universe.

2) This radiation is in the <u>microwave</u> part of the EM spectrum. It's known as the <u>cosmic microwave background radiation</u>.

3) For complicated reasons this background radiation is <u>strong evidence</u> for an <u>initial Big Bang</u>.

4) As the Universe <u>expands and cools</u>, this background radiation '<u>cools</u>' and <u>drops in frequency</u>.

There are microwaves in space — but not from ovens

I always thought the Universe was big enough already, but it's getting bigger all the time. How do we know? From clues like red-shift and microwave radiation — it'd be a bit difficult to go and measure.

The Origins of the Universe

Right now, all the galaxies are moving away from each other at great speed. But something must have got them going. That 'something' was probably a big explosion — so they called it the Big Bang...

It all *started off* with a very *big bang* (probably)

1) According to this theory, all the matter and energy in the Universe was compressed into a very small space. Then it exploded and started expanding.

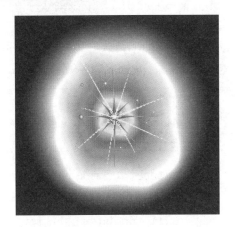

2) The expansion is still going on. We can use the current rate of expansion of the Universe to estimate its age. Our best guess is that the Big Bang happened about 14 billion years ago (though that might not be very accurate, as it's hard to tell how much the expansion has slowed down since the Big Bang).

3) The rate at which the expansion is slowing down is an important factor in deciding the future of the Universe. Without gravity the Universe would expand at the same rate forever. But as it is, all the masses in the Universe attract each other — and tend to slow the expansion down.

The *future* of the *Universe...*

1) The eventual fate of the Universe depends on how fast the galaxies are moving apart and how much total mass there is in it. But working this out is tricky — most of the mass appears to be invisible. Astronomers can only detect this dark matter by the way it affects the movement of things we can see.

2) Anyway, depending on how much mass there is, there are two ways the Universe could go:

Either a big crunch...

> If there's enough mass compared to how fast the galaxies are currently moving, the Universe will eventually stop expanding — and begin contracting. This would end in a Big Crunch.

...or just cold lonely oblivion

> If there's too little mass in the Universe to stop the expansion, it could expand forever with the Universe becoming more and more spread out into eternity.

Not everyone agrees but, the Big Bang seems to explain most evidence

Most scientists accept the idea of the Big Bang — it's the best way to explain the evidence we have at the moment. But if new evidence turns up, the theory could turn out to be rubbish. After all, there wasn't anyone around 14 billion years ago, taking photos and writing things down in a little notebook.

Warm-Up and Exam Questions

These warm-up questions are here to make sure you know the basics.
If there's anything you've forgotten, check up on the facts before you go any further.

Warm-up Questions

1) What type of star will form a black hole?
2) What is the 'cosmic microwave background'?
3) At the end of its main sequence stage, what does a star become?
4) What are the two possible fates for the Universe?
5) Our Sun is a second generation star. What does this mean?

Exam Questions

1   These sentences describe how a main sequence star is formed.
    They are in the wrong order.

    1.   The gravitational energy is converted into heat energy.

    2.   The star is stable as the forces from pressure and gravity are balanced.

    3.   A large cloud of dust and gas collapses under gravity.

    4.   Nuclear fusion begins.

    The correct order for these sentences is

    A   3   4   2   1

    B   1   2   3   4

    C   1   4   2   3

    D   3   1   4   2

    (1 mark)

2   Viewed from Earth, other galaxies seem to be

    A   moving away from us

    B   moving towards us

    C   contracting

    D   expanding

    (1 mark)

3   (a)   How do scientists believe the Universe started?

    (1 mark)

    (b)   Explain how the relationship between a galaxy's distance from Earth and its observed
          red shift suggests that the Universe is expanding.

    (2 marks)

    (c)   Scientists are unsure whether the Universe will expand forever or eventually begin to
          contract. Explain why.

    (1 mark)

Gravity, Mass and Weight

The only thing that stops you flying off the planet into space is gravity. You'd be <u>very lost</u> without it.

Gravity is the attraction between all masses

1) <u>Gravity</u> is what <u>attracts masses</u> to each other. <u>All masses</u> attract each other, but the <u>bigger the mass</u>, the <u>stronger its gravity</u> is.

2) The attraction is only noticeable when one of the masses is <u>really really big</u>, e.g. a planet or star. On Earth, gravity makes all things <u>accelerate towards the ground</u>, all at the same rate. Gravity also keeps <u>planets</u>, <u>moons</u> and <u>satellites</u> in their <u>orbits</u> around stars and planets.

The resultant force is the overall unbalanced force

The force of gravity acts on pretty much everything. But there's usually <u>more than one force</u> acting on any <u>object</u>, and if they <u>don't balance</u>, there'll be an <u>overall 'resultant' force</u>.

Unbalanced forces cause <u>objects</u> to <u>accelerate</u>. How it <u>accelerates</u> depends on its <u>mass</u> and the <u>force</u> — and Newton came up with a simple <u>formula</u> linking them:

$$\text{Resultant Force} = \text{Mass} \times \text{Acceleration}$$

$$F = ma \quad \text{or} \quad a = F/m$$

Make sure you use the right units:
<u>Force</u> in <u>newtons, N</u>;
<u>Mass</u> in <u>kg</u>;
<u>Acceleration</u> in <u>m/s^2</u>

Gravity keeps planets in orbit around stars

1) When a <u>planet</u> is <u>orbiting</u> a <u>star</u>, you can't see a force acting on it, but there must be one. The planet is <u>constantly accelerating</u> (its curved orbit means it's always <u>changing direction</u>, and a change of direction is an acceleration). And acceleration is always caused by a <u>force</u>.

2) Here, the force comes from the <u>gravitational attraction</u> between the <u>star</u> and planet.

Gravity, Mass and Weight

Weight and mass are not the same

1) <u>Mass</u> is the <u>amount of 'stuff'</u> in an object.

2) <u>Weight</u> is caused by the <u>pull of gravity</u>. An object has the <u>same mass</u> whether it's on <u>Earth</u> or on the <u>Moon</u>, but its <u>weight</u> will be <u>different</u> — since the Moon has <u>weaker</u> gravity (due to its smaller mass). This is why astronauts seem to be bouncing when they walk on the Moon.

3) <u>Weight</u> is a <u>force</u>, and has units of <u>newtons</u>. It's measured with a <u>spring balance</u> or <u>newton meter</u>.

4) <u>Mass</u> is <u>not</u> a force. It has units of <u>kilograms</u>.

Weight = mass × acceleration of free-fall

To work out the <u>weight</u> of an object, you'll need this equation:

$$W = m \times g$$
(in <u>N</u>)   (in <u>kg</u>)   (in <u>N/kg</u>)

Conveniently, a field strength of 1 N/kg causes an acceleration of 1 m/s², so <u>strength of gravity</u> always has the same value as <u>acceleration due to gravity</u> or <u>acceleration of free-fall</u> (measured in m/s²).

1) Hopefully it's obvious what W and m stand for.

2) The letter "g" represents the <u>strength of the gravity</u>, in N/kg. The value of g is <u>different</u> for <u>different planets</u>.

On <u>Earth</u> g = 10 N/kg. On the <u>Moon</u>, it's just 1.6 N/kg.

3) As you move away from the planet (or moon or whatever), the value of g decreases pretty quickly.

4) This formula is <u>hideously easy</u> to use:

<u>EXAMPLE:</u>   On the Earth, what is the weight, in newtons, of a 5 kg mass?

<u>ANSWER:</u>   W = 5 × 10 = <u>50 N</u>

Remember: gravity isn't the same on every planet

Apparently the whole point of Newton trying to explain gravity was to work out the <u>motion of the planets</u>. I always thought his theory of gravity just came out of nowhere and hit him on the head, so to speak.

Exploring the Solar System

If you want to know what it's like on another planet, you have three options — peer at it from a distance, get in a spaceship and go there yourself, or send a robot to have a peek...

Scientists explore the universe *from a distance*

The easiest way to find out about the Universe is to use <u>remote sensing</u> (i.e. looking from a <u>distance</u>). The idea is to detect <u>EM radiation</u>, but it's best if you can detect <u>different</u> kinds of EM wave...

- <u>Radio telescopes</u> are the <u>very large</u> dishes — they detect radio waves from space.

- <u>Optical telescopes</u> detect <u>visible light</u>.

- <u>X-ray telescopes</u> are a good way to 'see' violent, <u>high-temperature events</u> in space, like <u>exploding stars</u>. But they will <u>only</u> work <u>from space</u>, since the Earth's <u>atmosphere absorbs X-rays</u>.

Space exploration looks for *signs of life*

Part of the reason for exploring other planets it to try and find out if there's life out there.

1) <u>Sunlight</u> reflected from a planet, or refracted through its atmosphere, can give clues about what's on its <u>surface</u> or in its <u>atmosphere</u>.

2) Scientists look for <u>chemical changes</u> in a planet's atmosphere. Some changes in an atmosphere could be caused by things like volcanoes...

3) ...others are a <u>clue</u> that there might be life there (e.g. changing levels of <u>oxygen</u> and <u>carbon dioxide</u> in an atmosphere could be due to respiration or photosynthesis).

Little green men are the work of science-fiction — life on other planets is more likely to be simple organisms consisting of just a few cells.

Sometimes *manned spacecraft* are used

1) The Solar System is <u>big</u> — so big that even radio waves (which travel at 300 000 000 m/s) take several <u>hours</u> to cross it. Even from <u>Mars</u>, radio signals take at least a couple of <u>minutes</u>.

2) But sending a <u>manned spacecraft</u> to Mars would take at least a couple of <u>years</u> (for a round trip).

3) The spacecraft would need to carry a lot of <u>fuel</u>, making it <u>heavy</u> — and <u>expensive</u>.

4) And it would be difficult keeping the astronauts <u>alive</u> and <u>healthy</u> for all that time.

Exploring the Solar System

Keeping people *healthy* in space is *difficult*...

Getting people into space safely is difficult enough, but once they're up there, the astronauts need a lot of looking after...

1) The spacecraft would have to carry loads of <u>food</u>, <u>water</u> and <u>oxygen</u> (or be <u>very good</u> at <u>recycling</u>).

2) You'd need to regulate the <u>temperature</u> and remove <u>toxic gases</u> (e.g. CO_2) from the air.

3) The spacecraft would have to <u>shield</u> the astronauts from <u>cosmic rays</u> from the Sun.

4) Long periods in <u>low gravity</u> causes <u>muscle wastage</u> and loss of <u>bone tissue</u>.

5) Spending <u>ages</u> in a <u>tiny space</u>, with the <u>same people</u>, is psychologically <u>stressful</u>.

Space travel can be very stressful.

...sending *unmanned probes* is much *easier*

Build a <u>spacecraft</u>. Pack as many instruments on board as will fit. Launch. That's the basic idea here.

Advantages *of unmanned probes*

- They don't have to carry <u>food</u>, <u>water</u> and <u>oxygen</u> (or people) — so more <u>instruments</u> can be fitted in.

- They can withstand conditions that would be <u>lethal</u> to humans (e.g. extreme heat, cold or radiation levels).

- They're <u>cheaper</u> — they carry less, they don't have to come back to Earth, and less is spent on <u>safety</u>.

- If the probe does crash or burn up unexpectedly it's a bit <u>embarrassing</u>, and you've wasted lots of time and money, but at least <u>no one gets hurt</u>.

Disadvantages *of unmanned probes*

- Unmanned probes can't <u>think for themselves</u> (whereas people are very good at overcoming simple problems that could be disastrous).

- A spacecraft can't do maintenance and <u>repairs</u> — people can (as the astronauts on the Space Shuttle 'Discovery' had to do when its heat shield was damaged during take-off).

Make sure you know the pros and cons of unmanned probes

When people first sent things into space, we began cautiously — in October 1957, Russia sent a small aluminium sphere (Sputnik 1) into orbit around the Earth. A month later, off went Sputnik 2, carrying the very first earthling to leave the planet — a small and unfortunate <u>dog</u> called Laika.

Warm-Up and Exam Questions

You know the drill — check you can do the straightforward stuff with this warm-up,
then have a go at the exam questions below.

Warm-up Questions

1) Name the attractive force between masses.
2) Explain why a planet orbiting a star at a constant speed is still accelerating.
3) Give the unit of mass and the unit of weight.
4) Why can X-ray telescopes not be used on Earth?
5) What formula relates acceleration and force?

Exam Questions

1 Astronauts go to Mars and find that a 5 kg tool bag weighs 19 N.
  The strength of gravity on Mars must be

  A 0.26 N/kg

  B 3.8 N/kg

  C 10 N/kg

  D 95 N/kg

  (1 mark)

2 The diagram shows the forces acting on a firework rocket.

100 N

20 N

  The resultant force on the rocket is

  A 20 N

  B 120 N

  C 80 N

  D 100 N

  (1 mark)

3 Scientists can learn a lot by exploring the Solar System.

  (a) Give two advantages of sending unmanned probes rather than astronauts to explore
      the Solar System.

  (2 marks)

  (b) Give two disadvantages of using unmanned spacecraft rather than manned ones.

  (2 marks)

Revision Summary for Section Nine

And that's the end of Physics — but it's not quite over yet. You've got to check that what you think you learned actually stuck in your brain. The best way to do that is by answering all these lovely questions. There are a couple of formulas to learn to use in this section. Watch out for units though — before you bung your numbers into the formulas you have to make sure they're in the right units.

1) Sketch an atom, showing its protons, neutrons and electrons.

2) What is an unstable isotope?

3) Oxygen atoms contain 8 protons. What is the difference between oxygen-16 and oxygen-18?

4) Radioactive decay is a totally random process. Explain what this means.

5) List three sources of background radiation.

6) What is meant by ionisation?

7) Describe in detail the nature and properties of the three types of radiation: a, b and g.

8) What substances could be used to block: a) α-radiation   b) β-radiation   c) γ-radiation?

9) This data shows the count rate for a radioactive source at various times. Plot a graph of this data and use it to find the half-life of the substance.

Time (mins)	0	20	40	60	80	100	120
Count/minute	750	568	431	327	247	188	142

10) Explain how radiation damages the human body — a) at low doses, b) at high doses.

11) Explain which types of radiation are used, and why, in each of the following:
   a) medical tracers,   b) treating cancer,   c) detecting faults in aeroengine turbine blades,
   d) sterilisation,   e) smoke detectors,   f) thickness control.

12) What's the difference between a star and a planet?

13) What and where are asteroids? What and where are meteorites?

14) What does NEO stand for? What are they and why do scientists lose sleep over them?

15) Sketch the magnetic field of the Earth. What do scientists think causes the Earth's magnetic field?

16) Explain how solar flares can damage an artificial satellite.

17) What's the Milky Way?

18) What's a light year?

19) How big is the Universe?

20) Why are black holes 'black'? How can you spot one?

21) Describe the first stages of a star's formation. Where does the initial energy come from?

22) What happens inside a star to make it so hot?

23) What is a 'main sequence' star? How long does it last?

24) What are the final two stages of a small star's life? What are the two final stages of a big star's life?

25) Why will our Sun never form a black hole?

26) What's the main theory for the origin of the Universe? Give 2 important bits of evidence for it.

27) How long ago do we think the Universe began?

28) What are the two possible futures for the Universe, and what do they depend on?

29) What force keeps planets and satellites in their orbits?

30)*What force would be needed to give a 50 kg buggy an acceleration of 1.5 m/s²?

31)*I weigh 600 N on Earth. Find: a) my mass, and b) my weight on the Moon.

32) How could scientists investigate Neptune's atmosphere without actually sending someone there?

33) Describe 2 ways of looking for life on a planet without sending a spacecraft. Have they found any?

34) Explain 4 possible problems with going on a really long space journey.

* Answers on page 297.

Thinking in Exams

In the old days, it was enough to learn a whole bunch of <u>facts</u> while you were revising and just spew them onto the paper come exam day. If you knew the facts, you had a good chance of doing well, even if you didn't really <u>understand</u> what any of those facts actually meant. But those days are over. Rats.

Remember — *you might have to **think** during the **exam***

1) Nowadays, the examiners want you to be able to <u>apply</u> your scientific knowledge to newspaper articles you're reading or to situations you've <u>not met</u> before. Eeek.

2) The trick is <u>not</u> to <u>panic</u>. They're <u>not</u> expecting you to show Einstein-like levels of scientific insight (not usually, anyway).

3) They're just expecting you to use the science you <u>know</u> in an <u>unfamiliar setting</u> — and usually they'll give you some <u>extra info</u> too that you should use in your answer.

So to give you an idea of what to expect come exam-time, use the new <u>CGP Exam Simulator</u> (below). Read the article, and have a go at the questions. It's <u>guaranteed</u> to be just as much fun as the real thing.

Underlining or making notes of the main bits as you read is a good idea.

1. Blood glucose levels controlled by insulin.

2. Insulin added → liver removes glucose.

3. Not enough insulin → high blood glucose →death?

4. Carbohydrates cause problems for diabetics. So carbohydrates and glucose linked...

All cells need energy to function, and this energy is supplied by glucose carried in the blood. The level of glucose in the blood is controlled by the hormone insulin — if the <u>blood glucose level gets too high, insulin is introduced</u> into the bloodstream, which in turn <u>makes the liver remove glucose</u>.

Diabetes (type I) is where <u>not enough insulin is produced</u>, meaning that a person's <u>blood glucose level can rise</u> to a level that can <u>kill</u> them. The problem can be controlled in two ways:

a) Avoiding foods rich in carbohydrates. It can also be helpful to take exercise after eating <u>carbohydrates</u>.

b) Injecting insulin before meals (especially if high in carbohydrates).

<u>Questions:</u>

1. Why can it be helpful for a diabetic to take exercise after eating carbohydrates?

2. Suggest why a diabetic person should make sure they eat sensibly after injecting insulin.

3. Dave Edwards, a leading diabetes specialist, has made a new discovery about the condition. He decides to share his findings with the scientific community. Suggest two ways he could do this.

Clues — don't read unless you need a bit of a hand...

1. More complex carbohydrates are broken down to make glucose. What would normally happen if lots of glucose is suddenly put into the blood? How would this normally be controlled? And what happens in a diabetic?

2. Think about what insulin causes to happen.

3. This isn't a trick question — think of how you'd expect to read or hear about scientific discoveries.

Answers

1) Eating carbohydrates puts a lot of glucose into the blood.
Exercising can use up extra glucose, which helps stop blood glucose levels getting too high.

2) If they don't, blood glucose levels can drop dangerously low.

3) Any two sensible means of communication, e.g. conference, internet, book, journal, phone, meeting.

Answering Experiment Questions

Now then... in the exam there's every chance you'll get asked about an experiment. You might not have seen the experiment before (so don't panic if not) — you just need to keep your head and be scientific...

Graphs are used to show relationships

Melissa did an experiment on rate of reaction, mixing magnesium and excess dilute hydrochloric acid. She measured how much mass was 'lost' from the flask of reactants as a gas was given off. These were her results.

Time (s)	10	20	30	40	50	60	70	80	90	100
Loss in mass (g)	0.02	0.05	0.15	0.18	0.19	0.21	0.23	0.15	0.23	0.23

1. a) Eight of the points are plotted below. Plot the remaining **two** points.
   b) Draw the line or curve which best fits the points.

To plot the points, use a <u>sharp</u> pencil and make a <u>neat</u> little cross.

nice clear mark

smudged unclear marks

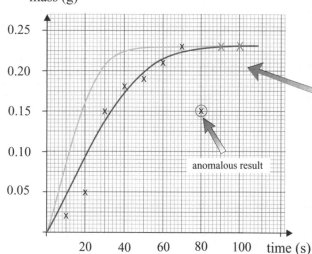

anomalous result

If your points lie roughly in a line, draw a line of best fit. If your points make a curve, draw a smooth curve instead. Whatever else you do, <u>don't</u> just join the dots.

When you're drawing your curve (or line), make it go as close to as many points as possible. It doesn't have to go <u>through</u> them — you want a <u>smooth</u> curve (or a <u>straight</u> line), not a wiggly one. In this case, the curve has to go through the <u>origin</u> (0, 0) as you know there'd be no gas given off if you hadn't mixed the reactants together yet.

2. Explain what your graph shows about how the rate of reaction changes with time.
   <u>The curve is steep for the first 30 s, showing
   that the reaction is quickest at first, but as
   the reactants are gradually used up, the
   reaction slows down and the curve flattens off.</u>

You might have some <u>anomalous results</u> — usually when you've done something daft, like reading a scale wrongly. You can <u>ignore</u> these anomalous results when you're drawing your curve (or line).

3. Sketch a graph to show the results you would expect to see if Melissa repeated the experiment using more concentrated acid, but keeping everything else the same.

You'd expect that, with more concentrated acid, the reaction would go <u>faster</u> — so the curve would be steeper at the beginning, and reach its highest value sooner. But there's the same mass of the other reactant (magnesium) as before, so the <u>same total amount</u> of gas will be given off — the final loss in mass will be the <u>same</u> as before.

Some things are best left unsaid — others are best ranted about...

Marks for <u>plotting points</u> (and <u>reading graphs</u>, for that matter) — it's the GCSE Science equivalent of <u>money for old rope</u>. Just do it carefully and check it when you've finished — you won't get easier marks.

Answering Experiment Questions

Not all experiments can be carefully controlled in a laboratory. Some have to be done in the real world. Unfortunately, this creates complications of its own.

Relationships do *NOT* always tell you the *cause*

Most car bodies are made from strong steel panels, but engineers are looking for innovative materials which will improve efficiency and safety. 'Alucars' has released a new car with body panels made of aluminium. The bar chart below shows how many accidents there were involving cars with aluminium bodies, and how many involving cars with steel bodies, in one year.

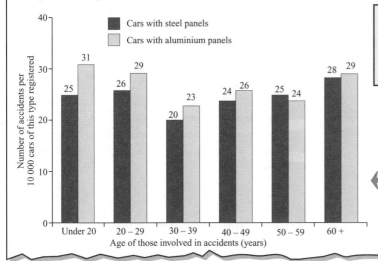

In <u>large</u> studies done outside a lab it's really <u>difficult</u> to keep all the <u>variables the same</u> and to make sure the <u>control group</u> are kept in the same conditions.

In this study the control group are the people in 'normal' cars, with steel panels.

This is a <u>bar chart</u>. It contains a <u>key</u> to tell you what colour bars relate to which group.

1. There are approximately 5000 aluminium cars registered in the county of Wessex. Use the bar chart to estimate how many under-20s will be involved in accidents in one year, in aluminium-panelled cars.
   31 ÷ 2 = 15.5 ≈ 16 people.

They're asking you the number of accidents you'd expect for 5000 cars — the graph tells you accidents per 10 000 cars. Don't get caught out — read the question really <u>carefully</u>. (And don't write something stupid with half a person in it.)

2. What conclusion can you draw from the results?
   There are proportionately more accidents involving cars with aluminium bodies than cars with steel bodies.

When <u>describing</u> the data and drawing <u>conclusions</u> it's really important that you don't say that having an aluminium-panelled car <u>causes</u> accidents. The graph only shows that there's a <u>positive relationship</u> between the two.

In studies like these where you're unable to control everything, it's possible a <u>third variable</u> is causing the relationship. E.g. aluminium-bodied cars would be <u>lighter</u> than steel-bodied cars, so they might appeal to people who like driving <u>fast</u>, and driving <u>faster</u> causes <u>more accidents</u>.

3. Suggest how the accident data may have been collected.
   e.g. from police records.

Use your <u>common sense</u> here.

Try to suggest a method to get <u>reliable</u> results. For example, it's very unlikely that the data would have been collected by a telephone survey or an internet search.

Revision causes Exam Success... (or does it — is there a third variable at work?)

Just looking at numbers (like here) doesn't mean you can say that one variable changing <u>causes</u> the other to change too. 'Ice cream sales' and 'cases of heatstroke' probably rise and fall together — but you can't say that ice cream sales <u>cause</u> heatstroke. (They'd more likely both be caused by a heatwave.)

Page 11

Warm-Up Questions

1) eyes, ears, nose, tongue, skin

2) It receives information from the sense organs and coordinates responses.

3) as electrical impulses

4) accommodation

5) The brain can use information from both eyes to judge distances, depth and how fast things are moving.

6) a synapse

Exam Questions

1 a) reflex action *(1 mark)*

   b) i) B *(1 mark)*

      ii) D *(1 mark)*

   c) When the electrical impulse reaches the end of the neurone, it stimulates the release of a chemical *(1 mark)*. The chemical diffuses across the gap/synapse to activate an electrical impulse in the next neurone *(1 mark)*.

   d) Any one of, they minimise damage to the body (because they are so quick) / they help to prevent injury *(1 mark)*.

2 a) E.g. when the receptors in the eye detect an increase in the level of light, they send impulses to the central nervous system (CNS) along sensory neurones *(1 mark)*. The CNS then sends impulses to the circular muscles along motor neurones *(1 mark)*, which causes the muscles to contract *(1 mark)*. This reduces the size of the pupil, so less light can enter the eye *(1 mark)*.
   Exactly the same reflex happens if the light is too dim, except that the CNS sends impulses to the radial muscles instead of the circular muscles. When the radial muscles contract, they increase the size of the pupil, so more light gets in.

   b) It prevents the eye's receptor cells from being damaged by very bright light *(1 mark)*.

Page 17

Warm-Up Questions

1) A chemical messenger that is carried in the blood and affects target cells.

2) E.g. hormonal responses are slower than nervous responses / hormonal responses persist for a longer time period than nervous responses / hormonal responses are normally quite widespread, whereas nervous responses are very localised.

3) the pancreas

4) Any four of, sperm production / voice deepening / enlargement of penis and testicles / growth of extra hair / development of muscles.

5) Hormones are given to a female to stimulate egg production. Eggs are collected and fertilised in a laboratory using the male's sperm. The eggs are grown into embryos, which are then transplanted into the female.

Exam Questions

1 C *(1 mark)*

2 B *(1 mark)*

3 a) It causes the lining to thicken and grow *(1 mark)*.

   b) It maintains the lining *(1 mark)*.

   c) day 14 *(1 mark)*

4 Any two of, e.g. abdominal pain / vomiting / dehydration / increased risk of cancer *(1 mark for each correct; maximum 2 marks)*.
   There are risks associated with most medical treatments, and IVF is no exception. People who decide to undergo IVF treatment should know and understand the risks, but if it's the only way they can have a child, then perhaps the benefits outweigh those risks.

Page 22

Warm-up Questions

1) The maintenance of a constant internal environment.

2) Any six of, e.g. carbon dioxide / oxygen / water / salt / ions / temperature / blood glucose levels.

3) In sweat, in breathing out, in urine and in faeces.

4) This is the optimal temperature for most enzymes in the body.

5) Genetically engineered microorganisms produce the human form of insulin so the insulin is not attacked by the immune system. It can be made in greater quantities.

Exam Questions

1 E.g. hairs stand on end, which traps a layer of warm air next to the skin *(1 mark)*. Sweat production decreases *(1 mark)*. Vasoconstriction occurs (blood vessels constrict to limit the amount of blood reaching the surface of the skin) *(1 mark)*. Shivering generates heat *(1 mark)*.

2 kidney *(1 mark)*

3 a) negative feedback *(1 mark)*

   b) i) pancreas *(1 mark)*

      ii) liver *(1 mark)*

      iii) glycogen *(1 mark)*

   c) i) E.g. after eating a carbohydrate-rich meal / eating sugary food *(1 mark)*.

      ii) cellular metabolism / respiration *(1 mark)*

      iii) vigorous exercise / being diabetic *(1 mark)*

   d) Injection of insulin *(1 mark)* and restriction of carbohydrate intake *(1 mark)*.

Page 23

Revision Summary for Section One

10)a) Response A

   b) Response B

Page 26

Warm-Up Questions

1) $C_6H_{12}O_6 + 6O_2 \rightarrow 6H_2O + 6CO_2$ (+ energy)

2) Any three of, e.g. build up larger molecules (e.g. proteins), muscle contraction, maintaining body temperature, active transport.

3) All living body cells.

4) haemoglobin

5) Systolic blood pressure is the blood pressure when the heart contracts. Diastolic blood pressure is the blood pressure when the heart relaxes. Systolic blood pressure is always higher than diastolic blood pressure.

Exam Questions

1 A — 4 *(1 mark)*

   B — 1 *(1 mark)*

   C — 3 *(1 mark)*

   D — 2 *(1 mark)*

2 a) glucose → lactic acid (+ energy) *(1 mark)*

   b) It allows her muscles to continue working when there is not enough oxygen *(1 mark)*.

   c) E.g. muscle fatigue, cramp/pain *(1 mark)*

   d) i) To provide extra oxygen needed to break down the lactic acid that has built up *(1 mark)*. To remove extra carbon dioxide *(1 mark)*.

      ii) To carry lactic acid to the liver where it is broken down *(1 mark)*. To carry extra carbon dioxide to the lungs *(1 mark)*.

Page 29

Warm-Up Questions

1) Chewing in the mouth / churning in the stomach.

2) They act as biological catalysts, breaking down large food molecules into smaller, more soluble ones.

) It lowers the pH in the stomach, giving the right conditions for the protease enzymes to work.

) amino acids

) by diffusion

Exam Questions

a) C *(1 mark)*

b) Any one of, bile is alkaline — it neutralises acid from the stomach to make conditions right for the enzymes in the small intestine to work / it emulsifies fats — giving a larger surface area for the lipase enzymes to work on. *(1 mark)*

a) carbohydrases *(1 mark)*

b) mouth *(1 mark)* and small intestine *(1 mark)*

a) i) lipases *(1 mark)*

ii) small intestine *(1 mark)*

iii) fatty acids and glycerol *(1 mark)*

b) They diffuse out of the gut into the lymphatic system *(1 mark)*.

Page 34

Warm-Up Questions

) Unbalanced diet, over eating and not enough exercise.

) E.g. to keep food moving smoothly through the digestive system / to avoid constipation.

) the liver

) Cholesterol deposits narrow the lumen of the arteries and can cause blood clots.

) Malnutrition is an imbalance of diet, starvation is the lack of food of any sort.

Exam Questions

B *(1 mark)*
Weight, temperature of surroundings and amount of exercise all have an effect on metabolic rate. Men usually have a higher metabolic rate than women.

a) i) $81 \div (1.85)^2 = 23.67$ *(1 mark)*
Don't forget you need to change the height from cm to m.

ii) normal *(1 mark)*

b) Any one of, e.g. arthritis / diabetes / high blood pressure / heart disease / some kinds of cancer *(1 mark)*.

a) E.g. protein-rich foods are often too expensive to buy *(1 mark)*.

b) E.g. growth, cell repair, cell replacement
(1 mark for each correct answer).

c) RDA (g) = 0.75 × 75 kg = 56.25 g *(1 mark)*

Page 41

Warm-up Questions

) A substance that alters chemical reactions in the body.

) E.g. where has the health claim been published? Was the research carried out by a qualified person? Was a large enough sample used? Have the results been backed up by other findings?

) A dummy treatment that doesn't contain any active drug.

) Any one of, e.g. caffeine, alcohol, nicotine.

) E.g. it can take many years to develop and test a drug, many potential drugs are rejected, scientists have to be paid.

Exam Questions

a) B *(1 mark)*

b) They slow down the activity of the nervous system *(1 mark)*.

a) Computer models, testing on human tissues, testing on live animals, clinical trials on human volunteers *(4 marks)*.

b) Double blind trials involve two groups of patients, one is given the drug, the other is given a placebo *(1 mark)*. Neither the patients or the scientists know which group is given the real drug and which is given the placebo until the results have been gathered *(1 mark)*. *This process prevents scientists or patients unconsciously affecting the results (it avoids bias).*

3 a) i) A physical or psychological need for a drug *(1 mark)*.

ii) The body has become used to having the drug so a higher dose is needed to get the same effect *(1 mark)*.

b) i) Class A *(1 mark)*.
In addition to being the most dangerous, Class A drugs also carry the most severe punishments if you're caught with them.

ii) E.g. heroin / LSD / ecstasy / cocaine *(1 mark)*.

c) Any one of, e.g. drug abuse can affect the immune system making infections more likely / sharing needles, which helps some infections to spread *(1 mark)*.

Page 46

Warm-up Questions

1) Any two of, e.g. the brain / lungs/breathing passages / liver

2) opiates

3) E.g. emphysema, bronchitis

4) Substances that can cause cancer.

5) E.g. because cannabis is illegal.

Exam Questions

1 a) Tar damages the cilia in the tubes of the lungs and windpipe. It also makes chest infections more likely *(1 mark)*.

b) Carbon monoxide reduces the oxygen carrying capacity of haemoglobin in the blood/red blood cells *(1 mark)*.

c) Nicotine is addictive *(1 mark)*.

2 a) i) Aspirin inhibits the formation of prostaglandins, the chemicals that cause swelling and sensitise the endings of nerves that register pain *(1 mark)*.

ii) Morphine interferes with the mechanism by which pain-sensing nerve cells transmit impulses. They also act on the brain to stop it sensing the pain *(1 mark)*.

b) Taking an overdose of paracetamol can cause liver damage *(1 mark)*.

3 a) Alcohol reduces the activity of the nervous system, making reactions slower *(1 mark)*. It can also lead to impaired judgement, poor balance and coordination and false confidence *(1 mark)*.

b) E.g. damage to brain cells/reduction in brain function / liver disease/ damage / increased risk of stroke/heart attack *(1 mark)*.

c) Any two of, e.g. increased crime/violence / costs to the NHS / costs to economy through lost working days.
(1 mark for each correct; maximum 2 marks)

Pages 53-54

Warm-Up Questions

1) A disease-causing organism.

2) They invade a cell and use the cell's DNA to make many copies of themselves.

3) Phagocytosis/to engulf and digest microorganisms, produce antibodies, produce antitoxins.

4) They live in or on the host obtaining nourishment from the host and give nothing in return.

5) Unique molecules which are present on the surface of cells/pathogens/ microorganisms.

Exam Questions

1 a) Active immunity is where the immune system makes its own antibodies after being stimulated by a pathogen or vaccination *(1 mark)*.

288

b) Passive immunity is where the body uses antibodies made by another organism (it is only temporary) *(1 mark)*.

2   A — 3 *(1 mark)*

B — 4 *(1 mark)*

C — 2 *(1 mark)*

D — 1 *(1 mark)*

3 a) Flu is caused by a virus *(1 mark)* and antibiotics are not effective against viruses *(1 mark)*.

b) Inappropriate use of antibiotics increases the chances of antibiotic-resistant strains of bacteria emerging *(1 mark)*.

4 a) Vaccinations help to prevent the outbreak of a disease in the first place because some people are immune *(1 mark)*. If an outbreak of the disease does occur vaccines help to slow down and stop the spread — if people don't catch the disease they cannot pass it on *(1 mark)*.

b) 4, 3, 1, 5, 2 *(1 mark)*

c) Any one of, e.g. swelling/redness at site of injection / feeling unwell *(1 mark)*

5   A — 3 *(1 mark)*

B — 4 *(1 mark)*

C — 2 *(1 mark)*

D — 1 *(1 mark)*

Page 55

Revision Summary for Section Two

6)   Professional runner, mechanic, secretary

20) a) 6 pm

b) 8 pm

c) No

Pages 62-63

Warm-Up Questions

1) the nucleus

2) mitosis

3) Short sections of a chromosome that determine the characteristics of an organism.

4) Any four of, e.g. moisture, temperature, mineral content of soil, sunlight, carbon dioxide.

5) The offspring receive genetic material from both parents. Meiosis also produces genetic variation in the gametes.

Exam Questions

1   D *(1 mark)*

2   C *(1 mark)*

3   D *(1 mark)*

4   C *(1 mark)*

5 a) Uncontrolled division of cells *(1 mark)*.

b) A permanent change in genetic material/gene/DNA *(1 mark)*.

c) Any two of, e.g. nuclear radiation / X-rays / UV light *(2 marks)*.

d) They may cause favourable changes *(1 mark)* which give the organism a survival advantage *(1 mark)*.
Not all mutations cause disease — some can cause favourable changes, some will have no effect at all. This is important for understanding natural selection (later on in this section).

6 a) Because they have exactly the same genes *(1 mark)*

b) Andrew and Peter look slightly different because their environment/ upbringing has affected their development/appearance *(1 mark)*.

c) i) Any one of, e.g. eye colour / gender / natural hair colour / blood group *(1 mark)*.

ii) Any one of, e.g. language / scars *(1 mark)*.

Page 65

Top Tip

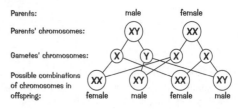

Pages 68-69

Warm-Up Questions

1) Versions of a gene.

2) Faulty alleles.

3) E.g. thick, sticky mucus is produced in the airways, pancreas and gut.

4) E.g. haemophilia, colour blindness.

5) No — some genes make people more likely to develop breast cancer, but this doesn't mean that they will develop breast cancer.

Exam Questions

1 a)

		parents' alleles	
		H	h
parents' alleles	H	HH	Hh
	h	Hh	h h

(1 mark)

b) A *(1 mark)*

2   B *(1 mark)*
There are loads of ethical issues concerned with genetic disorders and testing. Make sure you know about the ethical issues surrounding testing a foetus as well as an adult.

3   D *(1 mark)*

4 a)

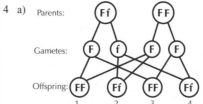

(1 mark for gametes correct, 1 mark for offspring correct)

b) They will all be unaffected *(1 mark)*.

c) i) 1 in 2 / 50% *(1 mark)*

ii) 2 and 4 *(1 mark)*

d) E.g. they may not want to risk passing on a genetic disorder to their children / may choose to avoid having children if they are a carrier *(1 mark)*.

Pages 76-77

Warm-Up Questions

1) All the DNA in the human body.

2) A human embryo that has been selected for its desired genes or genetically engineered to have desired genes.

3) Inserting a gene from one organism into the DNA of another.

4) An organism with exactly the same genetic material as another organism.

5) By using tissue culture and by taking cuttings.

Exam Questions

1 a) A *(1 mark)*

b) Life insurance could become impossible to get for someone who has a genetic likelihood of serious disease *(1 mark)*.

c) It may be possible to figure out what a suspect looks like from DNA found at the scene of a crime *(1 mark)*.

D *(1 mark)*

a) There are fewer alleles in a population *(1 mark)*.

b) E.g. if a population are all closely related and a new disease appears, all the population could be wiped out *(1 mark)*.

a) To produce lots of identical animals with desirable characteristics. *(1 mark)*

b) Sperm cells are taken from a male and egg cells are taken from a female *(1 mark)*. The sperm is used to fertilise the egg *(1 mark)*. The embryo that develops is split many times (to form clones) *(1 mark)*. The embryos are implanted into surrogate mothers *(1 mark)*.

a) The altering of genes *(1 mark)* in an organism to alter its characteristics *(1 mark)*.

b) E.g. to cut the gene out of the donor organism's chromosome *(1 mark)*. To cut the DNA of the recipient organism's chromosome *(1 mark)*. To join together the bacterial and human DNA *(1 mark)*.

c) Any one of, e.g. to give resistance to viruses / to give resistance to herbicides / to produce long-life fruit/vegetables / to give increased yields / to give crops with added nutrients *(1 mark)*.

d) Any one of, e.g. to produce milk containing human antibodies/ proteins / to produce low-cholesterol milk / to produce leaner meat / to give increased yields of wool from sheep *(1 mark)*.

e) Any two of, e.g. they could affect the numbers of other plants around the crop, reducing biodiversity / they could increase the risk of food allergies / they might not be safe / transplanted genes could transfer to other plants / super-weeds could develop. *(2 marks)*
Make sure you can explain the pros and cons of genetic engineering because it's a really controversial issue — and it could easily come up in the exam.

a) The nucleus was removed from a sheep egg cell *(1 mark)*. A complete set of chromosomes from an adult body cell (from a sheep) was inserted into the empty egg cell *(1 mark)*. This grew into an embryo which was implanted into a surrogate mother *(1 mark)*.

b) i) A cloned embryo that is genetically identical to the sufferer could be produced *(1 mark)* and embryonic stem cells extracted from it (to grow new cells or organs) *(1 mark)*.

ii) E.g. some people think it's unethical because embryos are destroyed *(1 mark)*.

Page 80

Warm-Up Questions

1) In the chloroplasts.

2) Chlorophyll

3) carbon dioxide + water $\xrightarrow[\text{chlorophyll}]{\text{light}}$ glucose + oxygen

$$6CO_2 + 6H_2O \xrightarrow[\text{chlorophyll}]{\text{light}} C_6H_{12}O_6 + 6O_2$$

4) The rate of photosynthesis increases with the availability of carbon dioxide, up to a point.

Exam Questions

C *(1 mark)*
Being small as well as soluble makes it a suitable transport molecule — it can diffuse in and out of cells really easily.

a) The rate of photosynthesis increases as light intensity increases *(1 mark)*.

b) Carbon dioxide *(1 mark)* or temperature *(1 mark)* has become a limiting factor.

c) Any one of, e.g. respiration / making cell walls / making proteins / making DNA *(1 mark)*.

d) It is insoluble *(1 mark)* so it doesn't cause cells to swell by causing water to enter *(1 mark)*.

Pages 87-88

Warm-Up Questions

1) A square frame used for counting organisms in a specific area.

2) A natural ecosystem is one where humans do not control the processes within it. An artificial ecosystem is one where humans deliberately promote the growth of certain living organisms and get rid of others which threaten their well-being.

3) Any one of, e.g. amount of food, amount of water, availability of shelter, level of competition, disease.

4) Sunlight

5) You can produce more food because you are reducing the number of stages in the food chain.

Exam Questions

1   B *(1 mark)*.
A mutualistic relationship is one where both organisms benefit, unlike parasitism (A & C) which only benefits one organism.

2   A *(1 mark)*

3   a) C *(1 mark)*

b) $30 \div 5 = 6$, $22 \times 6 = 132$ dandelions
(1 mark for correct answer, 1 mark for correct working)

4   a) D *(1 mark)*.

b) E.g. there was a decrease in the number of mice, which means less food for the owls *(1 mark)*.

c) It will probably decrease due to increased competition (between the owl and the other species) for food *(1 mark)*.

5   a) The mass of living material *(1 mark)*.

b) Any two of, energy losses occur between each stage of the chain / energy is lost as heat from respiration and used by the animals to move / not all of the organisms are eaten or digested *(1 mark for each correct answer, maximum 2 marks)*.

c) Because their body temperature is normally higher than their surroundings, so energy is lost as heat *(1 mark)*.

d) E.g. droppings / decay. *(1 mark)*

e) Trophic levels *(1 mark)*.

Page 96

Warm-Up Questions

1) How long ago the organism existed. Clues to what the organism looked like.

2) Any three of, e.g. large surface area compared to volume / produce small amounts of concentrated urine / they produce very little sweat.

3) Organisms with the most useful characteristics will survive to reproduce and pass these characteristics on to the next generation.

4) Evolution is the gradual change/adaptation of a population of organisms over time. Natural selection is the process by which evolution can occur.

Exam Questions

1   C *(1 mark)*

2   A — 2 *(1 mark)*
    B — 3 *(1 mark)*
    C — 4 *(1 mark)*
    D — 1 *(1 mark)*

3   a) Any one of, e.g. all organisms produce more offspring than could possibly survive but population numbers tend to remain fairly constant over long periods of time. / Organism in a species show wide variation in characteristics, some of which are passed on to the next generation. (1 *mark*)

b) There is variation within a population. Some individuals are better adapted to their environment *(1 mark)*. These individuals will be more likely to survive and reproduce *(1 mark)*, passing on the characteristics to the next generation *(1 mark)*. Over generations, the characteristics that increased survival becomes more common in the population *(1 mark)*.
Remember, individuals in a species are naturally selected for but individuals cannot evolve, only a species as a whole can evolve.

c) E.g. he could not explain how characteristics could be inherited / it went against many people's religious beliefs *(1 mark)*.

Page 97

Revision Summary for Section Three

10)
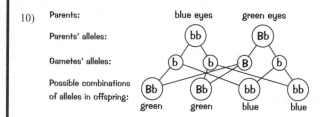

Pages 103-104

Warm-Up Questions

1) proton — positive / +1

neutron — neutral / 0

electron — negative / -1

2) protons and electrons

3) Mass number is the sum of the number of protons and the number of neutrons. Atomic number is the number of protons (or electrons) in an atom.

4) E.g. copper, iron (any solid element).

5) $C_6H_{12}O_6 + 6O_2 \rightarrow 6CO_2 + 6H_2O$

6)

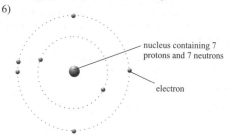

nucleus containing 7 protons and 7 neutrons

electron

7) The $_2$ in H_2SO_4 refers to 2 atoms (of H), while the 2 in 2NaOH refers to 2 molecules of NaOH.

Exam Questions

1   B *(1 mark)*

2   D *(1 mark)*

3   B *(1 mark)*

4 a) In a compound, different types of atoms are bonded together chemically. In a mixture they are not. *(1 mark)*

  b) i) noble gases / group 0 *(1 mark)*

     ii) 40 *(1 mark)*

  c) e.g. carbon dioxide / methane *(1 mark)*

  d) $2Mg + O_2 \rightarrow 2MgO$ *(1 mark for correct products and reactants, 1 mark for correctly balancing the equation)*

5   C *(1 mark)*

6 a) sulphuric acid + ammonia → ammonium sulphate *(1 mark)*

  b) $H_2SO_4 + 2NH_3 \rightarrow (NH_4)_2SO_4$ *(1 mark for correct products and reactants, 1 mark for correctly balancing the equation)*

  c) 15 *(1 mark)*
    There are eight atoms of hydrogen, one atom of sulphur, four atoms of oxygen, and two atoms of nitrogen.

Page 108

Warm-Up Questions

1) it increases

2) The elements go from gas to liquid to solid (at room temperature).

3) Any two of, e.g. neon — lasers or lights / argon — lasers or as an inert atmosphere in light bulbs / helium — balloons or airships.

4) hydrogen

5) E.g. silver bromide, zinc chloride (any metal halide).

Exam Questions

1 a) They are unreactive. *(1 mark)*

  b) They have a full outer shell of electrons. *(1 mark)*
    Having an outer shell that isn't full makes elements reactive.

  c) Argon *(1 mark)* — it provides an inert atmosphere *(1 mark)* which stops the filament burning away/oxidising *(1 mark)*.

2 a) Fluorine — gas *(1 mark)*

     Chlorine — gas *(1 mark)*

     Bromine — liquid *(1 mark)*

     Iodine — solid *(1 mark)*

  b) Arrow should be pointing upwards. *(1 mark)*

  c) i) displacement *(1 mark)*
      Chlorine is displacing iodine.

     ii) iodine/I/I_2 *(1 mark)*

Pages 116-117

Warm-Up Questions

1) Any two of e.g. iron, zinc, copper (any two transition metals).

2) Any three of, hard / strong / good electrical conductors / good conductors of heat / malleable / ductile / flexible.

3) metallic bonding

4) a metal ore

5) Any one of, zinc / iron / tin / copper (any metal below carbon in the reactivity series).

6) A mixture of metals, or a mixture of a metal and a non-metal, e.g. bronze — sculpture, medals / cupronickel — coins / solder — joining wires.

Exam Questions

1 a) E.g. potassium, sodium, calcium, magnesium, aluminium (any metal above carbon in the reactivity series). *(1 mark)*

  b) i) removal of oxygen (accept gain of electrons) *(1 mark)*

     ii) zinc oxide + carbon → zinc + carbon dioxide *(1 mark)*

  c) $Fe_2O_{3(s)} + 3CO_{(g)} \rightarrow 2Fe_{(s)} + 3CO_{2(g)}$ *(1 mark for correct products and reactants, 1 mark for correctly balancing the equation, 1 mark for correct state symbols)*

  d) C *(1 mark)*
    Zinc is more reactive than copper, so zinc will displace copper.

2   D *(1 mark)*

3   B *(1 mark)*

4 a)

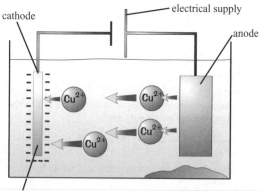

cathode

electrical supply

anode

pure copper deposit

Accept 'cathode' and 'pure copper deposit' either way round. (1 mark for each correct label)

  b) i) $Cu^{2+} + 2e^- \rightarrow Cu$ *(1 mark for correct ions and product, one mark for correctly balancing the equation)*

     ii) Cu^{2+} ions *(1 mark)* clearly moving towards the cathode *(1 mark)* — see diagram above. They move this way because they are positive and so are attracted to the negative cathode *(1 mark)*.

c) The supply of copper is limited/finite *(1 mark)*. Recycling uses less energy than mining and extracting new copper *(1 mark)* and so is cheaper *(1 mark)*. Recycling helps to conserve fossil fuels and so to reduce carbon dioxide emissions *(1 mark)*.

Page 118

Revision Summary for Section Four

7) 2 sodium atoms, 1 carbon atom and 3 oxygen atoms

8) a) bottom-left

b) top-right

c) top-left

d) bottom-right

10) calcium

11)a) $CaCO_3 + 2HCl \rightarrow CaCl_2 + H_2O + CO_2$

b) $Ca + 2H_2O \rightarrow Ca(OH)_2 + H_2$

Page 123

Warm-Up Questions

1) Any three of, e.g. as a building material / making glass / making cement / making slaked lime.

2) Any one of, e.g. destroys habitats / uses land / causes noise / causes pollution / leads to unsightly tips.

3) Cement is limestone that has been heated with clay. Mortar is cement mixed with sand and water.

4) Any two of, e.g. argon / carbon dioxide / water vapour / methane.

5) It melts ice and provides grip.

Exam Questions

1   A — 3 *(1 mark)*

B — 1 *(1 mark)*

C — 2 *(1 mark)*

D — 4 *(1 mark)*

2 a) $CaCO_3 \rightarrow CaO + CO_2$ *(1 mark)*

b) i) calcium hydroxide *(1 mark)*, $Ca(OH)_2$ *(1 mark)*

ii) e.g. neutralising acid soils *(1 mark)*

3 a) The gases have different boiling points *(1 mark)*.

b) Any two of, e.g. to make ammonia / to provide an unreactive atmosphere / to keep things cold (as liquid nitrogen). *(1 mark for each correct answer)*

Pages 132-133

Warm-Up Questions

1) Compounds made from carbon and hydrogen *only*.

2) Any three of, e.g. LPG / petrol / naphtha / kerosene / diesel / oil / bitumen.

3) They contain carbon-carbon double bonds.

4) Long-chain hydrocarbons are cracked to make more useful products / because there's more demand for short-chain fractions.

5) High temperature and a catalyst.

6) Any three of, e.g. transport / electricity generation / making plastics / heating / making medicines / making paints and dyes.

Exam Questions

1   A *(1 mark)*

2 a) i) There should be an M in the bottom box *(1 mark)*.

ii) There should be a B in the top box *(1 mark)*.
Fractions with bigger molecules have a higher boiling point, so condense at the higher temperatures at the bottom of the column. Fractions with smaller molecules have a lower boiling point, so don't condense until they reach the top of the column.

b) Any one of, e.g. jet engine fuel, domestic heating, paint solvent *(1 mark)*.

c) The explanation should contain three of the following points:
- the fractions have different boiling points
- the crude oil is heated
- the fractions boil
- fractions condense at different heights in the fractionating column
- fractions are tapped off where they condense
(1 mark per point; maximum 3 marks)

3   C *(1 mark)*

4   C *(1 mark)*

5 a) i) $2C_2H_6 + 7O_2 \rightarrow 6H_2O + 4CO_2$
(1 mark for correct products and reactants, 1 mark for correctly balancing the equation)
This is a bit of a tricky one — don't panic, just take one element at a time and keep pencilling in numbers till it all balances.

ii) Test its boiling point: 100 °C **OR** dip dry blue cobalt chloride paper in it: it turns pink **OR** add to white anhydrous copper sulphate crystals: they turn blue. *(1 mark)*. The limewater goes cloudy, indicating the presence of CO_2 *(1 mark)*.

b) i) An insufficient oxygen supply *(1 mark)*.

ii) carbon monoxide *(1 mark)* and carbon *(1 mark)*

6   A *(1 mark)*

Pages 139-140

Warm-Up Questions

1) volcano / mountain chain
Earthquakes occur at plate boundaries but they aren't a geological feature.

2) A (theoretical) single land mass / supercontinent made from the all present continents joined together.

3) By water vapour from volcanoes condensing.

4) radioactive decay

5) Any one of, e.g. bulging of the ground near the volcano / mini-earthquakes.

6) Any one of, e.g. it allowed complex organisms to evolve / it produced the ozone layer.

Exam Questions

1   A — 3 *(1 mark)*

B — 2 *(1 mark)*

C — 4 *(1 mark)*

D — 1 *(1 mark)*

2 a) There have been large variations in temperature and CO_2 concentration over the last 250 000 years *(1 mark)*. There is a positive correlation between CO_2 concentration and temperature *(1 mark)*.
The question's worth two marks, so you have to make two points.

b) The X should be drawn at 25 000 years ago *(1 mark)*.

3   C *(1 mark)*

4   B *(1 mark)*

5 a) The diagram should be labelled:

A – crust *(1 mark)*

B – mantle *(1 mark)*

C – inner core *(1 mark)*

D – outer core *(1 mark)*

b) nickel *(1 mark)*, iron *(1 mark)*

6 a) i) Any two of, Antarctica / the Arctic / Australia
(1 mark for each correct answer)

ii) They break it down *(1 mark)*.

iii) Any one of, e.g. aerosol propellants / coolants *(1 mark)*.

b) The answer should contain three of the following points:
- ozone protects against harmful UV radiation
- ozone levels have fallen
- incidence of skin cancer has increased
- but other factors, e.g. more holidays, may play a part

(1 mark per point; maximum 3 marks)
If the question says 'discuss' try to put across more than one point of view.

Pages 149-150

Warm-Up Questions

1) Any one of, e.g. improved sanitation / modern medicine / improved farming methods.

2) We are using up finite resources (crude oil, metals, etc.) more quickly and adding more and more pollution to our environment.

3) It absorbs heat radiated from Earth, preventing some of the heat being lost into space.

4) Any two of, e.g. lakes become acidic / kills fish and trees / damages limestone buildings/statues.

5) Any two of, e.g. nitrogen oxides / sulphur dioxide / carbon monoxide / water vapour.

6) The only product when it burns in air is water.

7) Any four of, e.g. energy value / availability / cost / ease of storage / toxicity / how polluting it is.

Exam Questions

1   C *(1 mark)*

2 a) The diagram should be labelled:

A – respiration *(1 mark)*

B – compounds in plants *(1 mark)*

C – photosynthesis *(1 mark)*

D – burning *(1 mark)*

b) Large scale deforestation increases the amount of CO_2 in the atmosphere *(1 mark)* because: CO_2 is released into the atmosphere when trees are burnt to clear land, microorganisms feeding on dead trees release CO_2 through respiration, fewer trees mean that less CO_2 is absorbed from the atmosphere in photosynthesis *(2 marks for any two of these points)*.

3   C *(1 mark)*

4   A — 3 *(1 mark)*

B — 1 *(1 mark)*

C — 4 *(1 mark)*

D — 2 *(1 mark)*

5 a) i) methane and CO_2 *(1 mark)*

ii) By microorganisms digesting waste *(1 mark)*.
It can be burnt to heat water/generate electricity *(1 mark)*.

iii) Advantage: any one of, e.g. cheap / carbon neutral / renewable / uses waste products *(1 mark)*.

Disadvantage: e.g. production is slow in cold weather *(1 mark)*.

b) i) by electrolysing water *(1 mark)*

ii) It's highly explosive, so it has to be kept in very secure containers *(1 mark)*.

iii) Any one of, e.g. you would need a specially designed, expensive engine / hydrogen isn't widely and cheaply available / large, strong gas containers are heavy and would increase fuel consumption *(1 mark)*.

Page 154

Warm-Up Questions

1) Any two of, e.g. fossil fuels (oil/natural gas/coal) / uranium / stone (e.g. limestone) / metals (e.g. aluminium).

2) Any one of, e.g. collecting / sorting / processing / transporting material.

3) Any two of, e.g. reduces amount of waste / less energy is used for mining/transport/processing / uses less resources / can be cheaper.

Exam Questions

1 a) D *(1 mark)*

b) i) Loss of a species from an ecosystem unbalances it, which can lead to the extinction of other species *(1 mark)*.

ii) Many useful products (e.g. medicine, food) come from plants and animals. If the organisms that produce them are extinct, we can't use these products *(1 mark)*.

2   A *(1 mark)*

3 a) indicator species *(1 mark)*

b) i) Any one of, e.g. some species of lichen / mayfly larvae *(1 mark)*

ii) Answer to this depends on the answer to part i).
One of:

- the level of sulphur dioxide in the air *(1 mark)* because the lichens grow better where there is less sulphur dioxide pollution *(1 mark)*

- how clean / well oxygenated water is *(1 mark)* because mayfly larvae thrive where water is clean/contains plenty of oxygen *(1 mark)*

There are lots of other possible answers to this question — you just need to make sure that your answers to parts i) and ii) match.

Page 155

Revision Summary for Section Five

13)   E.g. $2C_2H_6 + 5O_2 \rightarrow CO_2 + 6H_2O + C + 2CO$ (+ energy)

Page 159

Warm-Up Questions

1) Monomers are small molecules which can be joined together to give much larger molecules called polymers.

2) The higher the melting point, the stronger the forces holding the polymer chains together.

3) covalent bonds

4) E.g. for making kettles.

Exam Questions

1)   biodegradable, burnt, toxic, recycle, expensive *(5 marks)*
The majority of polymers aren't biodegradable, and this has significant environmental consequences. Chemists are currently working on biodegradable polymers though.

2) a)

(1 mark)

b)

(1 mark)

c) i) Any one of, e.g. window frames/CDs *(1 mark)*

ii) Any one of, e.g. clothing/synthetic leather *(1 mark)*

Page 165

Warm-Up Questions

1) solvent, binding medium, pigment

2) To make sure they are safe to use.

3) Pigments that absorb light, store the energy in their molecules and gradually release it as light. They could be used for, e.g. signs, 'glow in the dark' hands on watches.

4) Oil-based paints have something that dissolve oil as the solvent. Water-based paints have water as the solvent.

5) E.g. sunglasses that become darker in more intense sunlight.

4) Oil-based paints have something that dissolve oil as the solvent. Water-based paints have water as the solvent.

5) E.g. sunglasses that become darker in more intense sunlight.

Exam Questions

1 a) A colloid is a mixture of tiny particles of one kind dispersed (but not dissolved) in another substance. *(1 mark)*

b) Because the dispersed particles are so small that they don't settle at the bottom. *(1 mark)*

c) i) Gloss paint, because it is harder-wearing. *(1 mark)*

ii) Emulsion paint, because it dries quickly and produces only low levels of harmful fumes. *(1 mark)*
You don't need to memorise all the different properties, just be able to interpret information about them.

2 a) esterification *(1 mark)*

b) carboxylic acid + alcohol → ester + water *(1 mark)*

c) Any five of, evaporate easily, non-toxic, don't react with water, don't irritate the skin, insoluble in water, have a pleasant smell.
(1 mark for each correct property, maximum of 5 marks)

Page 172

Warm-Up Questions

1) E.g. it kills microbes / destroys the poisons present in some raw foods.

2) The protein molecules change shape irreversibly when heated (this is known as denaturing).

3) Antioxidants are chemicals that stop fat and oil reacting with oxygen.

4) The fruits or seeds are crushed. The oil is separated from the crushed plant material (by a centrifuge or using solvents). The oil is then distilled to refine it.

5) Unsaturated oils contain double bonds between some of the carbon atoms in their carbon chains. Saturated fats contain no double bonds.

6) A nickel catalyst, at about 60°C.

7) It makes the food more suitable for people with diabetes or people who are dieting.

Exam Questions

1 C *(1 mark)*
Potatoes are plants, so each cell is surrounded by a cellulose cell wall, which humans can't digest.

2 Extract the colour from the food sample by placing it in a test tube with a few drops of solvent *(1 mark)*. Put a spot of the coloured solution on a pencil baseline on some filter paper *(1 mark)*. Stand the filter paper in a beaker with some solvent. *(1 mark)*

3 a) hydrophilic —●⌇⌇⌇⌇— hydrophobic *(1 mark)*

b) Hydrophobic means that it does not mix with water (it doesn't 'like' water) *(1 mark)*. Hydrophilic means that it is attracted to water (it 'likes' water) *(1 mark)*.

c)

(1 mark)

d) They prevent the mayonnaise emulsion separating into its component liquids / they keep the oil and water mixed well together. *(1 mark)*

Page 176

Warm-Up Questions

1) A fuel made from vegetable oils. It is a renewable resource / gives off less sulphur dioxide pollution when burnt.

2) ethanol

3) An enzyme in yeast.

4) (10%) ethanol mixed with (90%) petrol. It is used extensively in Brazil.

5) Distillation produces more concentrated alcohol.

Exam Questions

1 A — 4 *(1 mark)*
B — 2 *(1 mark)*
C — 3 *(1 mark)*
D — 1 *(1 mark)*

2 a) Crude oil is non-renewable. As it runs out it will become more expensive *(1 mark)*.

b) i) $C_2H_5OH + 3O_2 \rightarrow 2CO_2 + 3H_2O$ *(1 mark for correct reactants and products, 1 mark for correctly balancing the equation)*

ii) Ethanol can be produced by fermenting crops (e.g. sugar cane) *(1 mark)*. The crops absorb carbon dioxide as they grow (by photosynthesis) *(1 mark)*.
Photosynthesis takes carbon dioxide out of the air and produces oxygen.

Pages 182-183

Warm-Up Questions

1) One in which water is a reactant.

2) Because water is a product in esterification reactions.

3) An alkali is a soluble base.

4) Acids — 0 to 6/6.9, alkalis — 7.1/8 to 14, neutral — 7

5) dehydration

6) iron + oxygen + water → hydrated iron(III) oxide

Exam Questions

1 A — 2 *(1 mark)*
B — 1 *(1 mark)*
C — 4 *(1 mark)*
D — 3 *(1 mark)*

2 a) It is added to a solution of the unknown substance *(1 mark)*. A precipitate often forms *(1 mark)* and its colour indicates specific metals *(1 mark)*.

b) i) oxygen: relights a glowing splint *(1 mark)*

ii) ammonia: turns damp red litmus paper blue *(1 mark)*

iii) hydrogen: gives a squeaky pop when a lighted splint is held to it *(1 mark)*

iv) chlorine: bleaches damp litmus paper (turns it white) *(1 mark)*

3 B *(1 mark)*
This is the only reaction between an acid and a base and the only one in which the products are a salt and water.

4 B *(1 mark)*

5 B *(1 mark)*

6 a) thermal decomposition *(1 mark)*

b) i) $NaHCO_3$ *(1 mark)*

ii) sodium hydrogencarbonate → sodium carbonate + carbon dioxide + water *(1 mark)*

iii) Bubble the gas that is produced into limewater *(1 mark)*. You would expect the limewater to go cloudy/milky. *(1 mark)*

iv) Baking powder *(1 mark)* — to make cakes/bread rise *(1 mark)*.

Pages 190-191

Warm-Up Questions

1) Exothermic — gives out energy/heat to surroundings.
Endothermic — takes in energy/heat from surroundings.

2) E.g. burning, neutralisation

3) Measuring the volume of gas evolved at regular time intervals.
Measuring the loss of mass from the reaction mixture at regular time intervals.

4) How often and how hard particles collide / the frequency of collisions and the energy of the colliding particles.

5) The minimum energy needed for a reaction to happen.

6) By heating / raising the temperature of the reactants. This gives the particles more (kinetic) energy so they move faster.

Exam Questions

1   A — 3 *(1 mark)*

    B — 2 *(1 mark)*

    C — 1 *(1 mark)*

    D — 4 *(1 mark)*

2   A *(1 mark)*

3   C *(1 mark)*

4 a) i)   The particles are free to move past each other *(1 mark)* but there is some force of attraction between them so they tend to stick together *(1 mark)*.

      ii) The particles are free to move / have virtually no force of attraction between them *(1 mark)* so they move in straight lines until they collide with each other or with the sides of the container *(1 mark)*.

      iii) The particles have strong forces of attraction between them / are not free to move *(1 mark)*, so they stay in a regular arrangement *(1 mark)*.

   b) i)   Particles at the surface of a liquid overcome the forces of attraction from other particles and escape *(1 mark)*.

      ii) Perfumes have to evaporate easily so they can reach our smell receptors quickly *(1 mark)*.

5 a) i)   The reaction is finished/the reactants have been used up. *(1 mark)*
      When the line flattens out no more products are being formed. This means that the reaction has finished.

      ii) X at initial steep section. *(1 mark)*

      iii)

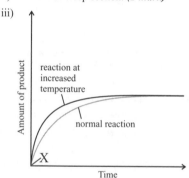

      The curve should have a steeper initial section *(1 mark)* and level off at the same level *(1 mark)*.
      Increasing the temperature of a reaction increases the rate of reaction.

   b)   There are more particles per unit volume *(1 mark)* therefore more (successful) collisions per unit time *(1 mark)*.

Page 192

Revision Summary for Section Six

29)   When using the concentrated acid it will take less time to produce the same amount of gas than when using the dilute acid — the rate of reaction is faster. The slope of the graph (time vs volume of gas) will be steeper for the acid which produces the faster rate of reaction.

Page 197

Warm-Up Questions

1) A type of energy / a measure of energy.

2) When there is a temperature difference between two places.

3) Breaking the bonds between molecules to turn the liquid into a gas.

4) The amount of energy needed to melt 1 kg of a substance without changing its temperature.

Exam Questions

1   B *(1 mark)*
    18 000 (energy needed) ÷ 80 (temperature change) ÷ 0.5 (mass of iron) = 450 J/kg/°C

2   A — 3 *(1 mark)*

    B — 2 *(1 mark)*

    C — 4 *(1 mark)*

    D — 1 *(1 mark)*

3 a) $2000 \times 60 \times 2 = 240\,000$ J *(2 marks, allow 1 mark for correct working)*

   b) Mass = energy ÷ specific latent heat
      $= 240\,000 \div 2\,260\,000$
      $= 0.106$ kg $(= 106$ g$)$
      (2 marks, allow 1 mark for correct working)

Page 203

Warm-Up Questions

1) Particles that vibrate faster than others pass on their extra kinetic energy to their neighbours.

2) Heated air expands, so it becomes less dense than the surrounding cooler air and rises.

3) infrared (radiation)

4) E.g. make the surface darker in colour, make the surface rougher.

5) Any three of, e.g. loft insulation / cavity wall insulation / draught proofing / double glazing / using thick curtains.

Exam Questions

1   C *(1 mark)*

2   D *(1 mark)*
    Light and shiny surfaces are best for reflecting heat radiation.

3 a) $300 \times 0.25 = £75$ *(1 mark)*

   b) $300 - 255 = £45$ saved per year *(1 mark)*.
      Payback time = cost ÷ saving per year = $350 \div 45 = 7.8$ years *(1 mark)*.

Page 207

Warm-Up Questions

1) chemical energy

2) Energy cannot be created or destroyed, only converted from one form to another.

3) More of the input energy is transformed into useful energy in modern appliances.

4) E.g. wind-up radio, clockwork toy.

5) Some energy is always wasted and so all the input energy isn't transformed usefully.

Exam Questions

1   B *(1 mark)*
    Efficiency = Useful energy output ÷ Total energy input.

2   D *(1 mark)*

3 a) $1200 - 20 - 100 = 1080$ J *(1 mark)*
      Energy cannot be destroyed or created, so the total energy output must equal the energy input.

   b) By reducing the amount of energy wasted as sound / by having a quieter motor *(1 mark)*.
      Only the energy converted to sound is wasted — kinetic and heat energy are what you want from a hairdryer.

295

Pages 218-219

Warm-Up Questions

1) Any two of, e.g. it releases greenhouse gases/contributes to global warming / it causes acid rain / coal mining damages the landscape.

2) Nuclear power stations produce radioactive waste, which is dangerous and difficult to dispose of / risk of a catastrophic accident.

3) Organic matter that can be burnt to release energy.

4) Pumped storage is a method of storing electricity whereas hydroelectric power schemes actually generate electricity.

5) Any two of, e.g. they're expensive to install / are inefficient / can only generate electricity when there is enough sunlight.

Exam Questions

1 a) Heat energy from inside the Earth *(1 mark)*.
  b) The source of energy will never run out *(1 mark)*.
  c) Because there are no hot rocks near the surface of the Earth in the U.K. *(1 mark)*

2   D *(1 mark)*
    Tides are caused by the pull of the Sun and Moon's gravity.

3   A — 3 *(1 mark)*
    B — 4 *(1 mark)*
    C — 1 *(1 mark)*
    D — 2 *(1 mark)*

4 a) 2 000 000 ÷ 4000 = 500 *(1 mark)*
  b) If the wind isn't blowing strongly, the turbines will not generate as much as 4000 W each *(1 mark)*.
  c) Any two of, e.g. they believe it would spoil the view (visual pollution) / cause noise pollution / kill or disturb local wildlife *(1 mark for each correct answer)*.

5   A *(1 mark)*
    Wind, solar and wave power all depend on the weather, which is changeable.

6 a) i) by (thermal) radiation *(1 mark)*
    ii) by conduction through the metal pipe *(1 mark)*
    iii) by convection currents in the water *(1 mark)*
  b) Because black surfaces are good absorbers of heat radiation *(1 mark)*.

7 a) Any two of, e.g. wave / tidal / geothermal / biomass *(1 mark for each)*.
  b) Any two of, e.g. set-up time / set-up costs / running costs / impact on environment / social impact *(1 mark for each)*.

Page 220

Revision Summary for Section Seven

2)   900 J/kg/°C
4)   791 kJ
10) a) 2 years,
    b) light bulb B
16)   0.7 (or 70%)

Pages 228-229

Warm-Up Questions

1) The flow of electrons/charge round a circuit.

2) Alternating current (AC) changes direction whereas direct current (DC) always flows in the same direction.

3) It increases.

4) The creation of a voltage (and maybe current) in a wire which is experiencing a change in magnetic field.

5) The network of cables and pylons that distributes electricity across the country.

6) voltage = current × resistance / V=IR / resistance = voltage ÷ current / current = voltage ÷ resistance

Exam Questions

1   A *(1 mark)* $V = I \times R = 2.5 \times 10 = 25V$

2   A — 4 *(1 mark)*
    B — 1 *(1 mark)*
    C — 2 *(1 mark)*
    D — 3 *(1 mark)*

3   D *(1 mark)*

4 a) Because electrical current is generated. *(1 mark)*
  b) Any one of, e.g. move the magnet out of the coil, move the coil away from the magnet / insert the south pole of the magnet into the same end of the coil / insert the south pole of the magnet into the other end of the coil. *(1 mark)*
  c) Any one of, e.g. push the magnet into the coil more quickly / use a stronger magnet / add more turns to the coil. *(1 mark)*
  d) Zero / no reading. *(1 mark)* *Movement is needed to generate a current.*

5 a) Chemical energy *(1 mark)*
  b) i) 25 000 × 1000 = 25 000 000 kJ = **25 000 MJ** *(1 mark)*
     ii) (10 000 ÷ 25 000) × 100 = **40%** *(1 mark)*

Page 231

Top Tip

1)   Power = voltage × current = 230 × 12 = 2760 W = 2.76 kW
2)   Time = energy ÷ power = 0.5 ÷ 2.76 = 0.181 h = 11 min

Page 234

Warm-Up Questions

1) The rate of transfer of electrical energy.

2) kilowatt hours (kWh)

3) So that if there is a fault causing the live wire to touch the metal case, a large current flows to Earth and 'blows' the fuse, making the appliance safe.

4) If there is a fault, a large current will flow in the live wire, melting the fuse — this will break the circuit and isolate the appliance.

5) RCCBs can be reset whereas fuses need to be replaced each time there is a fault and they melt.

Exam Questions

1   C *(1 mark)* $I = P \div V = 1200 \div 230 = 5.22 A$

2   B *(1 mark)*

3 a) 8.75 kWh = 8.75 × 1000 × 60 × 60 = **31 500 000 J** *(1 mark)*
  b) 8.75 × 14.4 = **126p** (=£1.26) *(1 mark)*
  c) £1.26 × 90 = **£113.40** *(1 mark)*
  d) 12 × 8.75 = 105p, 126 – 105 = **21p** *(1 mark)*
     (OR 14.4 – 12 = 2.4, 2.4 × 8.75 = 21p)

Page 239

Warm-Up Questions

1) The number of complete wavelengths that pass a point within 1 second.

2) Interference occurs when waves with similar frequencies combine, so that both signals are distorted.

3) Diffraction

4) Any one of, e.g. sound waves / shock waves / seismic P-waves.

Exam Questions

1   C *(1 mark)*

2   A *(1 mark)* $\lambda = v \div f = 300\,000\,000 \div 100\,000\,000 = 3\,m$.
    Don't forget that you have to convert MHz to Hz before doing the calculation.

3 a) 5 cm *(1 mark)*
  b) 1 complete wave would pass a point every 2 seconds, so f = 1 ,
     2 = **0.5 Hz** *(1 mark)*

Answers

c) It will halve. *(1 mark)*
Because frequency and wavelength are inversely proportional.

Page 242

Warm-Up Questions

1) It ionises atoms/molecules within cells — this can cause e.g. mutations in DNA, or kill cells.

2) It causes heating / can cause burns.

3) Because they damage / cause thinning of the ozone layer (which protects us from UV radiation).

4) Ultraviolet

Exam Questions

1   C   *(1 mark)*
No one knows for sure whether microwave radiation from phones and masts is causing any health problems.

2   A — 3 *(1 mark)*

     B — 4 *(1 mark)*

     C — 2 *(1 mark)*

     D — 1 *(1 mark)*

3   a)   i)   Hazard: is absorbed by water molecules, causing heating which could damage living cells. *(1 mark)*

         Use: any one of, e.g. mobile phone signals/communication / cooking *(1 mark)*

     ii)   Hazard: any one of, e.g. it's highly ionising / it can cause cell damage/mutations in DNA / it can cause cancer / it can cause radiation sickness. *(1 mark)*

         Use: any one of, e.g. sterilising food/medical instruments / treating cancer / as a medical or industrial tracer. *(1 mark)*

   b)   Any one of, e.g. stay in the shade / wear sunscreen / keep covered up with clothing. *(1 mark)*

Pages 249-250

Warm-Up Questions

1) A wave enters a different medium (with its wavefronts at an angle to the boundary) and changes direction. This happens because the wave changes speed in the new medium.

2) Certain frequencies of microwave are absorbed by water molecules (which are present in all food). This heats the water and so heats the food.

3) When light is travelling from a denser medium towards a less dense medium and its angle of incidence at the boundary is greater than the critical angle.

4) A thin glass or plastic fibre which is used to transmit light/IR waves using repeated total internal reflections.

5) Because the light no longer hits the boundary within the optical fibre at a big enough angle, and so is not internally reflected (and transmitted along the fibre).

6) (short wavelength) radio waves

Exam Questions

1   A — 2 *(1 mark)*

     B — 3 *(1 mark)*

     C — 4 *(1 mark)*

     D — 1 *(1 mark)*

2   C *(1 mark)*

3   a)

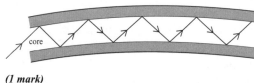

(1 mark)

---

   b)   The light ray meets the core/outer boundary at an angle greater than the critical angle, so is totally internally reflected, and this happens repeatedly. *(1 mark)*

4   a)   It acts like a camera, recording light reflected from the iris. (This image is then processed by computer into a code.) *(1 mark)*

   b)   Any one of, e.g. iris scanning is quick and easy / iris patterns don't normally change during a person's lifetime / there is very little chance of mistaking one iris for another. *(1 mark)*

   c)   Any one of, e.g. eye injuries or surgery can occasionally change the iris pattern / people feel that their personal freedom may be being threatened / iris pattern data is believed to be unfakeable and so genuine identity theft or mistakes may not be believed. *(1 mark)*

5   a)   Because X-rays mostly pass straight through soft tissue, so the structure of soft tissues doesn't appear on X-ray images. *(1 mark)*

   b)   E.g. leave the room while the X-ray image is being taken, wear a lead apron (if they need to remain in the room). *(2 marks)*

6   Advantage: any one of, e.g. doctors can check on the baby's development / parents can find out the sex of the baby. *(1 mark)*

     Disadvantage: any one of, e.g. scientists are not certain that ultrasound is completely safe / parents may be more likely to want an abortion if their baby is found to be the "wrong sex" or disabled. *(1 mark)*

Page 255

Warm-Up Questions

1) Digital signals can only take two values, on and off, whereas analogue signals can take any values within a given range.

2) P-waves

3) P-waves

4) They are refracted as the properties of the Earth's interior change.

5) Any one of, monitoring groundwater levels / detecting foreshocks / monitoring radon gas emissions / monitoring animal behaviour.

Exam Questions

1   D *(1 mark)*

2   A *(1 mark)*

3   a)   on or off (1 or 0) *(1 mark)*

   b)   Both types of signal will have lost energy and weakened. *(1 mark)*

   c)   Interference on digital signals can be much more easily filtered out by his radio. *(1 mark)*

Page 256

Revision Summary for Section Eight

4)   4.8 ohms

13)   279 kJ

14) a)   Units = power × time = 2.5 × (2/60) = 0.0833 = 0.08 kWh

   b)   0.0833 × 12 = 1 p

20)   150 m/s

27)   A and D (which are identical)

Page 259

Top Tip

     beta

Page 261

Warm-Up Questions

1) An unstable atom emits radiation and becomes more stable, often changing into a different element.

2) Atoms which have the same number of protons but different numbers of neutrons.

3) Alpha, beta and gamma

4) Gamma radiation

5) Any one of, e.g. cosmic rays, rocks, radon gas, food, building materials, human activity (e.g. fallout from nuclear bomb tests), living things.

Exam Questions

1   A — 4 *(1 mark)*

   B — 1 *(1 mark)*

   C — 3 *(1 mark)*

   D — 2 *(1 mark)*

2   C *(1 mark)*

3 a) The time taken for half of the unstable nuclei in a sample to decay / the time taken for the count rate to halve. *(1 mark)*

   b) one quarter / 25% *(1 mark)*

   c) beta *(1 mark)*

Page 266

Warm-Up Questions

1) A weak alpha source is used to ionise the air between two electrodes so that a current can flow. If the alpha radiation is absorbed by smoke, the current stops and the alarm sounds.

2) Because it is ionising and can damage cells.

3) Any two of, e.g. never look directly at the source / always handle a source with tongs / never allow the source to touch the skin / never have the source out of its lead-lined box for longer than necessary.

4) Any one of, e.g. wear lead-lined suits / work behind lead/concrete barriers / use robotic arms to do tasks in highly radioactive areas.

5) Any one of, e.g. medical tracers / treatment of cancer.

Exam Questions

1   A *(1 mark)*

2   D *(1 mark)*

3 a) Beta radiation is directed through the paper towards a detector *(1 mark)*. If the paper's thickness changes, so does the amount of radiation reaching the detector, and a control moves the paper rollers accordingly. *(1 mark)*

   b) Alpha radiation would be completely stopped by paper. *(1 mark)*

   c) Gamma radiation would not be blocked by even thick paper and so couldn't provide any information about thickness. *(1 mark)*

Page 272

Warm-Up Questions

1) Between Mars and Jupiter.

2) dust, gas and ice

3) Any one of, e.g. asteroids orbit in the same plane as the planets; comets do not / asteroids were formed at the same time as the planets; comets were not / comets have highly elliptical orbits; asteroids do not.

4) A very large group of stars/solar systems.

5) Because its gravity is so strong that light can't escape.

Exam Questions

1   A — 3

   B — 4

   C — 2

   D — 1   *(1 mark for each correct answer)*
   1 must be the comet as it orbits in a different plane from the planets.

2   C *(1 mark)*

3 a) Explosions on the Sun's surface that throw out huge amounts of energy and charged particles/cosmic rays. *(1 mark)*

   b) i) The Earth is shielded by its magnetic field. *(1 mark)*

      ii) The Aurora Borealis / northern lights. *(1 mark)*
      Cosmic rays pass on energy to particles in the atmosphere which then emit visible light.

c) i) Any one of, e.g. navigation, spying, weather forecasting, communications. *(1 mark)*

   ii) The charged particles can cause a surge of current which may damage satellites' electrical components. *(1 mark)*

Page 276

Warm-Up Questions

1) A very large/heavy star.

2) The low frequency microwave radiation that comes from all parts of the Universe.

3) A red giant.

4) It may contract and end in a 'Big Crunch' or it may expand forever.

5) Our Sun (and Solar System) formed from material created in the supernova of a previous star.

Exam Questions

1   D *(1 mark)*

2   A *(1 mark)*

3 a) All the matter and energy was in a very small space, then there was an explosion/a 'Big Bang'. *(1 mark)*

   b) The further the galaxy, the greater the red shift *(1 mark)*.
   This shows that the more distant the galaxy, the faster it's moving away from us. This must mean that the Universe is expanding. *(1 mark)*

   c) It depends on what the total mass of the Universe is, but we don't know how much 'dark matter' there is. *(1 mark)*

Page 281

Warm-Up Questions

1) gravity

2) Because it is constantly changing direction, so its velocity is changing.

3) Mass — kg. Weight — N (weight is a force).

4) Because the Earth's atmosphere absorbs most of the X-rays from space.

5) Force = mass × acceleration / a = F ÷ m / m = F ÷ a

Exam Questions

1   B *(1 mark)* F = ma, a = F ÷ m = 19 ÷ 5 = 3.8 N/kg

2   C *(1 mark)* 100 − 20 = 80 N

3 a) Any two of, e.g. they don't need to carry food/water/oxygen so more instruments can be taken / they can withstand conditions lethal to humans / they're cheaper. *(1 mark for each correct answer)*

   b) E.g. unmanned spacecraft can't think for themselves, unmanned spacecraft can't carry out repairs to the spacecraft or instruments. *(1 mark for each correct answer)*

Page 282

Revision Summary for Section Nine

30)   75 N

31)a) 60 kg

   b) 96 N

Index

A

absolute scale 193
AC (alternating current) 225, 227
acceleration 277, 278
accommodation 10
acid rain 142, 187, 208
acids 179
activation energy 189
adaptations 91, 92, 94, 95
addiction 37, 42
addition polymers 156
adult cell cloning 75
aerobic respiration 24
alcohol 37, 38, 42, 43, 174, 175
alkali metals 105
alkanes 125, 127
alkenes 126, 127, 156
alleles 58, 64, 66
alloys 114
alpha radiation 257, 258, 262, 264
alternative fuels 148
aluminium 114, 153
amino acids 78
ammeter 222
amplification 252
amplitude 235, 237, 251
amps 223
analogue signals 251
angle of incidence 244
angle of reflection 244
animal testing 39, 162
anode 113
anomalous results 284
antibiotics 37, 52
    resistance to 51, 91
antibodies 48, 49
antigens 48, 49
antiseptics 51
antitoxins 48
antiviral drugs 52
applying scientific knowledge 283
arctic conditions 94
argon 107
arteries 25
artificial ecosystems 81
artificial immunity 50
asexual reproduction 59
aspirin 45
asteroid belt 267
astronauts 278, 279
atmosphere 136-138, 143, 267, 279
atmospheric pollution 146
atoms 98, 102
atomic number 98
Aurora Borealis 270
average 3

B

background radiation 257, 270
    cosmic 274
bacteria 21, 42, 47, 48, 51, 70
balance 187
balancing equations 102
barbiturates 38
bases 58, 179

beta radiation 257, 258, 262, 264
Big Bang Theory 274
binding medium 160
binocular vision 10
biodegradable 173
biodegradable plastics 157
biodiesel 173
biodiversity 71, 81, 151
biogas 214
biomass 83, 214
black dwarf 273
black holes 271, 273
blood 27, 28
    cells 25
    clotting 48
    groups 56
    pressure 25
    sugar levels 20
    vessels 19
body mass index (BMI) 33
body temperature 18
boiling 186, 195, 196
brain 6, 38
brain damage 43, 44
breast milk 50
breathable material 158
broadcasts 252
bromine water 125
bronchitis 42
buffalo 86
burglar detectors 223
burning 184, 187

C

cacti 95
caffeine 37, 38
calcium carbonate 119
calcium hydroxide 119
camels 94
cancer 16, 32, 42, 47, 57, 61, 246, 262
    breast 66
    treatment 265
cannabis 36, 37, 45
capillaries 25
car airbag sensors 164
car bodies 109
carbohydrases 27
carbohydrates 20, 30
carbon cycle 144
carbon dating 260
carbon dioxide 18, 24, 78, 79, 121, 136, 138, 143, 279
carbon monoxide 42, 129, 145, 146
carboxylic acids 161
carcinogens 42
catalysts 146, 156, 174, 189
catalytic converters 146
cathode 113
cavity walls 202
CD players 248
cellulose 78, 166
cement 120
central heating 130, 194
Central Nervous System (CNS) 6
CFCs 138, 241
CGP Exam Simulator 283
charge, electric 257, 258
charged particles 270
chemical digestion 27

chemical energy 204
chemical industry 122
chemical reactions 102
chemical reaction rates 187, 188
chemical weathering 187
Chernobyl disaster 210
chest infections 42
chlorine 106, 122
chlorophyll 78
chlorophytum (spider plant) 59
cholesterol 33, 36, 170
chromatography 169
chromosomes 58-60
cigarette smoke 61
cilia 42, 48
ciliary muscles 10
classification 33
cleaner species 86
climate change 208
clinical trials 39
clones 59
cloning 74, 75
coal 208
coal-fired power stations 217
Collision Theory 189
colloids 160
colour blindness 47
combustion 144, 184, 185
comets 137, 268
competition 85, 91
complete combustion 128
compounds 99
computer models 39, 147
concentration 189
concrete 120
condensing 195
conduction 198, 246
consumers 83
continental crust 134
contraception 16
control group 285
convection 199, 246
convection currents 134, 199, 221, 269
conventional current 221
copper 113
core (of the Earth) 254
coronary heart disease 33
corrosion 115, 179
cosmetic testing 162
count rate 260
covalent bonds 125, 126
cracking 127, 184
critical angle 244
CRO (cathode ray oscilloscope) trace 221
crude oil 99, 124, 127, 130, 131, 173, 175
crust 134
current 221, 222-224, 227, 230, 233
cuttings 74
cystic fibrosis 66, 72

D

dark matter 275
Darwin, Charles 92
DC (direct current) 212, 225
decay, radioactive 257, 260
decommissioning 210
deficiency diseases 30, 47
deflection 258

dehydration 19
dehydration reactions 177
denaturing 166
density 243
depressants 44
desert conditions 94
designer babies 73
diabetes 21, 32, 47
diastolic pressure 25
diesel 173
diet 30, 32
diffraction 238
digestion 27
digital signals 251
dimmer switches 251
dinosaurs 91
discrimination 73
diseases 47, 51, 52, 57, 72, 91
displacement reactions 106
distillation 170
DNA 47, 58, 59, 72, 241
    fingerprint 72
Dolly the sheep 75
dominant alleles 64
double glazing 202
double insulation 233
Down's syndrome 247
draught-proofing 202
drugs 37-39
dyes 164
dynamos 224, 225

E

E-numbers 168
Earth 121, 267, 269
    structure 253
earth wire 232
earthquakes 134
earthworms 135
ecological niche 85
economic factors 5
ecosystems 81-83
ecstasy 38
effectors 6, 7
efficiency 205, 206
egg cells 14, 16, 60
electric
    currents 130, 214, 221, 269
    fields 258
electrical
    conductors 224
    disturbances 252
    energy 204, 224
    power 230, 231
    pressure 221
    safety devices 232, 233
    signals 248, 251
electrodes 264
electrolysis 111, 113, 114, 122
electrolytes 19
electromagnetic (EM) induction 224, 226, 270
electromagnetic radiation 235, 240, 241, 251, 274
electron pump 113
electrons 98, 221, 257, 258, 270
elements 99, 109
embryos 16, 75
embryo transplants 74
embryonic stem cells 75
emulsion 160, 168

Index

endothermic reaction 184
energy 24, 78, 84, 193, 195, 202, 204, 205, 209, 230, 231, 235, 240, 270
  conservation 204
  resources 208
  transfer 184, 185
environment 56, 82, 91, 208, 217
enzymes 27, 70, 79, 174
esters 161
ethanoic acid 161
ethanol 130, 174, 175
ethene 126, 156
ethics 4
evaporation 186
evidence 1, 3
evolution 90, 91
exercise 19, 20, 24
exothermic reactions 184
experiments 2, 285
exposure, radiation 262
extinction 91
eyes 9, 10, 48
eye colour 56

F
faeces 18
farming 142
  battery 83
fats 30
fatty acids 170
fertilisation 61
fertilisers 57
fertility, control of 16
fibre 30
fitness 31
flame tests 180
flavourings 168
flu 52
Flue Gas Desulphurisation (FGD) 146
fluorescent tubes 241
food 85, 209
  chains 83
  colourings 168, 169
  production 84
forces 278
forensic science 72
fossil fuels 124, 136, 138 145, 208, 209
fossils 89, 90, 135
fractional distillation 121, 124, 127
freezing 195
frequency 225, 235, 237, 240, 251, 274
FSH (follicle-stimulating hormone) 12, 15, 16
fuel 130, 131, 173
fungi 47
fuses 233

G
galaxies 271, 274, 275
gall bladder 28
gametes 60
gamma rays 257-259, 265, 270
gas-fired power stations 217
gases 186
gasohol 175
Geiger counter 260
gene therapy 67, 70

generators 211, 212, 224, 226
genes 56, 58, 64, 70, 72
genetic diagrams 64-67
genetic disorders 47, 66, 70, 72
genetic engineering 21, 70, 71
genetically modified (GM) crops 70
geology 260
geothermal energy 209, 211
glands 6
glass 120
global dimming 131, 146
global warming 131, 143, 210
gloss paint 160
glow-in-the-dark 164
glucose 20, 24, 28, 78, 84
glycerol 27
gravitational potential energy 204, 273
gravity 271, 273, 275, 277, 278
greenhouse effect 138, 143, 144, 208, 210
growth 56, 59, 78

H
Haber process 122
habitats 81, 85
haemoglobin 25, 42, 47
haemophilia 47, 66
hair colour 56
half-life 260
halogens 106
health claims 35
heart 25
  attack 42
  disease 32, 36, 57, 170
heat 94, 193-195, 205
heat radiation 200, 201
heat stroke 19
hepatitis 52
heterozygous 64
high blood pressure 32
HIV (human immunodeficiency virus) 38
homeostasis 18
homozygous 64
hormonal responses 13
hormones 12-16, 20
hot rocks 211
Human Genome Project 67, 72, 73
hybrids 82
hydration reactions 177
hydrocarbons 127, 129
hydroelectric power 209, 216
hydrogen 105, 122, 148
hydrogenation 171
hydrophilic 167
hydrophobic 167
hypothesis 1

I
ibuprofen 45
ice ages 137
identical twins 56
immune system 38
immunisation 49, 50
incomplete combustion 129, 146
indicator species 151
infections 47, 48
inflammatory response 48
infrared radiation 240, 247

inherited disorders 56
inner planets 267
insulation 198, 202
insulin 20, 21, 70
intelligence 57
interference 237, 252
internal combustion engines 145
intestine 27
ionisation 258, 262
ions 19, 113, 258, 270
iris reflex 9
iron 114, 134, 179, 187, 269
isotopes 257
IVF (in vitro fertilisation) 16, 73

J
joules 193
Jupiter 267

K
KEVLAR® 163
kidneys 19
kilowatts 230, 231
kinetic energy 198, 204, 224
kinetic theory 186
kwashiorkor 32

L
lactic acid 24
landfill 214
lasers 107, 240
latent heat 196
leguminous plants 86
leprosy 40
lichen 151
light 9, 78, 79, 85, 204
light-dependent resistor 223
limestone 119, 120
limewater 128, 181
lipases 27
lipids 78
lithium 105
lithosphere 134
litmus paper 181
live wire 232
liver 28, 33, 43, 44
long-chain molecules 156
longitudinal waves 253
lung cancer 42
lungs 18, 25, 44
lymphatic system 28
lysozyme 48

M
magnetic field 134, 224, 258, 269, 270
malnutrition 32
mammals 82, 84
mantle 134, 254
margarine 171
Mars 267, 279
mass 275, 277, 278
mass number 98
measles 49
melamine resin 157
melting 195, 196
Mendeleev, Dmitri 100
menstrual cycle 14, 15
Mercury 267
metabolism 20, 31
metals 109-115

metal halides 106
metal ores 111
metallic bonds 110
meteorites / meteors 267
methane 144
microbes 47-49
microwaves 240, 245, 246
Milky Way 271
mineral deficiency 47
minerals 30, 85, 111
mitosis 59, 60
mixtures 99
money for old rope 284
monomers 156
monounsatured fats 171
mortar 120
motor neurones 6, 7
MRSA (methicillin-resistant Staphylococcus aureus) 51
mucus 42, 48
multiple pregnancies 16
mumps 49
muscles 6, 7, 9, 31, 38
mutations 61, 262
mutualistic relationships 86

N
nanomaterials 163
national grid 226, 227
natural gas 208
natural selection 61, 92
Near Earth Objects (NEOs) 268
negative feedback 18
neon 107
Neptune 267
nervous responses 13
nervous system 6, 38, 43, 44
neurones 6, 9, 44
neutral wire 232
neutralisation reactions 179
neutron star 273
neutrons 98, 257
newtons 277
nickel 134
nicotine 38, 42
night-vision equipment 247
nitinol 164
nitrates 78
nitric acid 145
nitrogen 121, 136
nitrogen oxides 145
nitrogen-fixing bacteria 86
noble gases 107
noise 252
non-destructive testing 265
non-renewable energy resources 208
northern lights 270
nuclear
  energy 204, 208, 210, 211
  fusion 273
  radiation 61, 258, 262-265
nucleus 58, 98, 257, 260

O
obesity 32
ocean 136
oestrogen 12, 14-16
ohms 223
oil 208
oil spills 131

Index

oil-based paints 160
oil-fired power stations 217
opiates 45
outer planets 267
ovaries 12
oxidation 179
oxygen 18, 24, 47, 78, 121, 144, 279
oxygen debt 24
ozone 136, 138, 145, 241

P

P waves 253
painkillers 44, 45
paint 160
pancreas 12, 21
Pangaea 135
paracetamol 44
parasites 47, 86
particulates 173
passive immunity 50
pathogens 47, 48
payback time 202
penetration 258
peppered moths 91
perfumes 161, 162, 186
periodic table 100, 101, 109
personal freedom 248
pH 28, 179
phagocytes 48
photocells 212
photosynthesis 78, 79, 136, 144
physical barriers 48
physical digestion 27
pigments 160
pituitary gland 12
placebo 40
planets 267, 271, 278
plants 56, 57, 78, 79
plant oils 170, 171
plasma 25, 28, 48
plastics 156, 157
platelets 25
plate tectonics 135
plutonium 208, 211
poisons 91, 95
polar bears 94
polio 50
pollution 142, 145, 146, 152, 173
polymerisation 156
polymers 126, 156, 157, 160
population size 85
Post-it® Notes 163
power 230
power stations 130, 145, 214, 217, 226, 227
precautionary principle 147
precipitates 180
predators 85, 91, 95
predisposition 66
prenatal scanning 247
prescription drugs 37
preservatives 168
pressure 189
prey 10, 85
primary consumers 83
principle of the conservation of energy 204
processed foods 168
products 102, 187

progesterone 16
propene 126, 156
prostaglandins 45
proteins 30, 32, 48, 58, 78, 166
protons 98, 257
protozoa 47
providers 83
puberty 14
pupil 9
pyramids of biomass 83

Q

quadrat 81
quantitatively 83
quarrying 120, 142

R

radial muscles 9
radiation sickness 262
radiators 199
radio waves 245
radioactivity 257, 258
  radioactive decay 134
  radioactive waste 210, 211
rainforests 151
rate of reaction 187, 189
reactants 102, 187
receptors 6, 9, 38, 45
recessive alleles 64
recommended daily allowance (RDA) 30
recycling 113, 152, 153
red blood cells 47
red giant 273
red-green colour blindness 47, 66
reduced gene pool 74
reduction with carbon 114
reflection of waves 247
reflex arc 7
reflexes 7
refraction 238, 243, 244, 254
renewable energy resources 208, 209
reproduction 14
  asexual 59
  sexual 60
residual current circuit breaker (RCCB) 232
resistance 221-223
respiration 24, 25, 28, 78, 84, 144
  aerobic 24
  anaerobic 24
respiratory system 48
resultant force 277
retina 9
rock salt 121
rocks 90, 111
rubella 49
rust 179, 187

S

S waves 253
salt 33, 106
sample size 81
satellites 245, 268, 270
saturated hydrocarbons 125
screening 73
scurvy 47
secondary consumers 83

secondary sexual characteristics 14
seeds 78
seismic waves 253
Semmelweiss, Ignaz 51
sensory neurones 6, 7
sex 14
sexually transmitted infections (STIs) 16
shock waves 235
skin 18, 48
small intestine 27
smallpox 50
smart alloy 115
smoking 42, 43
sodium 19, 105
solar energy 212, 213
solar flares 269, 270
Solar System 267, 268, 271, 279, 280
solvents 44, 160, 170
sound waves 235
species 82
specific heat capacity 193
specific latent heat 196
sperm cells 16, 60
sports drinks 19
stabilisers 168
stable isotopes 257
starch 78
stars 273
starvation 30, 32
states of matter 186
statins 36
sterilising 265
stimulus 6, 7
stomach 27, 28
strength of gravity 278
sugar 20
sulphur dioxide 142, 145, 146, 151
Sun 200, 241, 271
Sun Protection Factor (SPF) 241
sunburn 241
sunscreen 163, 241
supernovas 273
surface area 25, 94, 188
survival of the fittest 92
suspensory ligaments 10
sweating 18, 19, 196
sweeteners 168
synapses 8
systolic pressure 25

T

tar 42
target cells 12
tectonic plate 134, 135
Teflon® 163
telescopes 268
temperature 19, 79, 196, 198, 247, 273
terrorists 211
testes 12
testosterone 14
thalidomide 40
thermal decomposition 119, 127, 178, 184
thermal energy 204
thermistors 223
thermograms 202
thermometers 251
thickness gauges 264
tidal power 215

tissue culture 74
totally indisputable fact 2
toxins 47
trans fats 171
transition metals 109
transverse waves 253
trophic levels 83
tumour 47
turbines 212, 224, 226

U

ultrasound 235, 247
ultraviolet (UV) light 61, 236, 241
Universe 274
unsaturated hydrocarbons 126
unstable isotopes 257
urine 18, 19

V

variation 56, 57, 60, 92
  environmental 57
  genetic 56, 60
vegetable oil 170, 173
veins 25
viruses 47, 52
visible light 240, 279
vitamin C 47
vitamin deficiency 47
vitamin E 170
volcanoes 134, 136, 137
voltage 221-224, 230

W

warfarin 91
water 18, 24, 78, 85, 94, 177
water pollution 142
water-based paints 160
watts 230
wave energy 215
wavelength 235, 238
Wegener, Alfred 135
weight 277, 278
white blood cells 48
white dwarf 273
wind power 212
wireless communication 238
woolly mammoths 91

X

X-rays 61, 270, 271, 279

Y

yeast 174

Z

zygote 61